Principles and Applications of

Laboratory Instrumentation

Principles and Applications of

Laboratory Instrumentation

Sheshadri Narayanan, PhD
Clinical Professor
Department of Pathology
New York Medical College
New York, NY

ASCP Press
American Society of Clinical Pathologists • Chicago

Notice
Trade names and equipment and supplies described herein are included as suggestions only. In no way does their inclusion constitute an endorsement or preference by the American Society of Clinical Pathologists. The ASCP did not test the equipment, supplies, or procedures and, therefore, urges all readers to read and follow all manufacturers' instructions and package insert warnings concerning the proper and safe use of products.

Library of Congress Cataloging-in-Publication Data

Narayanan, Sheshadri, 1936–
Principles and applications of laboratory instrumentation /
Sheshadri Narayanan.
p. 232
ISBN 0-89189-273-7
1. Chemistry, Clinical—Instruments—Handbooks, manuals, etc.
2. Biochemistry—Instruments—Handbooks, manuals, etc. I. Title.
[DNLM: 1. Equipment and Supplies. 2. Laboratories. QY 26 N218p]
RB40.N29 1989
616.07'56—dc20
DNLM/DLC
for Library of Congress 89-17595

Printed in the United States of America.

93 92 91 90 89 5 4 3 2 1

Contents

Figures

Acknowledgement

The support given me over the years by Dr Joseph G. Fink, Clinical Associate Professor of Pathology, New York Medical College, and Dr Victor Tchertkoff, Professor and Acting Chairman, Department of Pathology, New York Medical College, and Chief, Department of Pathology and Director of Laboratories at Metropolitan Hospital Center, New York, was in large measure responsible for the preparation of this book. To Dr Desmond Burke, Professor of Pathology and Director of Laboratories, at University Hospital at State University of New York at Stony Brook, I am greatly indebted for encouragement. I would like to thank the publishers Appleton-Century-Crofts for permission to use some of the material which appeared previously in the *Journal of Clinical Laboratory Automation.* The painstaking secretarial support given to me principally by Ms Carol Perello and also by Ms Lisa Scialla made it possible for completion of this book. Finally, I am grateful for the invaluable support of the staff at the ASCP Press.

1
Fundamental Concepts of Spectrophotometry

Spectrophotometry is the cornerstone of analytical instrumentation. In the past decade, automated analyzers have emerged—such as computerized continuous flow analyzers, discrete rate analyzers, and certain immunoassay equipment—that utilize spectrophotometric measurements. A basic understanding of spectrophotometric theory is a prerequisite to understanding modern analytical methods. In this chapter, basic concepts, such as the origin of spectra and the electromagnetic spectrum, will be explored before the discussion of the design of photometers and spectrophotometers.

Spectra

Origin

Electromagnetic radiation can be visualized as waves with electric and magnetic components that oscillate in planes perpendicular to each other. When electromagnetic radiation, specifically the electric component, interacts with matter (molecules), the radiation behaves as particles and not as waves. Energy absorbed or emitted by a molecule is related to the changes in its energy state. Absorption of electromagnetic energy by a molecule results in a change in the energy levels of outer shell or valence electrons, and gives rise to electronic spectra. Molecular spectra arising from absorption of electromagnetic energy in the ultraviolet or visible region are caused by electronic transition (outer shell electrons) and changes in the vibration and rotational energy levels of the molecule.

The Electromagnetic Spectrum

The electromagnetic spectrum ranges from highly energetic cosmic rays to electric current. Electromagnetic radiation powerful enough to cause changes in nuclei will result in a spectrum in the gamma ray region. Changes in the inner shell electrons provide spectra in the x-ray region. Radiation that effects changes in the rotational and vibrational energy states results in infrared spectra. Orientation of the spin of an unpaired electron in a magnetic field provides electron spin resonance (ESR) spectra. Likewise, orientation of certain nuclei in a magnetic field yields nuclear magnetic resonance (NMR) spectra. Instrumentation that will resolve spectra in the ultraviolet and the visible region of the electromagnetic spectrum are those most commonly used in the clinical laboratory.

Concept of Wavelength

As we noted earlier, electromagnetic energy can be visualized in the form of waves. The distance between waves measured either from crest to crest or trough to trough is referred to as the *wavelength* (λ). Energy itself is expressed in discrete quantities called quanta or photons, which travel at the speed of light. The speed of light (c) is equal to the product of wavelength and frequency $\lambda\nu = c/n$, where c is the speed of light in a vacuum (3×10^{10} cm/sec), which must be divided by the refractive index of the medium (n). The relationship between the energy of a photon (E) and the frequency of electromagnetic radiation is expressed by the equation $E = h\nu$, where h is Planck's constant, 6.624×10^{-27} erg-sec, ν is frequency in cycles per second or hertz (Hz), and E is the energy of the photon. As we see the energy of photons is proportional to the frequency of electromagnetic radiation.

Wavelength is inversely proportional to energy. The shorter the wavelength, the greater the number of photons compressed over that distance; thus, the short wavelength region will be one of high energy. Conversely, the longer wavelengths correspond to regions of lower energy. Thus, a wavelength of 340 nm, which is in the ultraviolet (UV) region, has a higher energy than one of 500 nm, which is in the visible region.

Nature of Absorption Spectra

The wavelength at which a compound absorbs light is related to its molecular structure. Absorption in the UV region is related to energy transitions in the valence electrons of the molecule. Saturated compounds yield spectra in the far UV region (below 200 nm). Functional groups, such as the carbonyl group and carbon-carbon double and triple bonds, have maximum absorbance in the UV region and are referred to as *chromophores.* However, when chromophores are separated by conjugated double bonds they have more resonance energy. Consequently, less energy is required for electronic transitions, and such compounds absorb in the visible region of the spectrum. Since UV spectra result from electronic transitions, any change in charge caused by a change in pH, temperature, salt concentration, or change of solvent will shift the spectra.

The color of a solution that is visible to the naked eye corresponds to the light that the solution does not absorb; its complementary color is the color that is absorbed. The wavelength at which a colored solution absorbs is the one to be chosen for making absorption measurements. Thus, a blue-colored solution transmits blue light and absorbs yellow light. Such a solution should be measured at a wavelength corresponding to 580 to 595 nm, the range in which yellow is absorbed. In contrast, a yellow-colored solution is measured in the range where blue is absorbed (435 to 480 nm), and a yellowish green solution in the range where violet color is absorbed (380 to 435 nm).

Laws of Absorption Spectroscopy

Bouguer and Lambert observed that transmittance of light through absorbing material (contained in a cuvette in spectrophotometric analyzers) decreases exponentially with an increase in the path length or thickness of the cuvette. Beer and Bernard noted that the absorbance of a solution is directly related to its concentration in dilute solutions (less than 10^{-2} M). A combination of the findings of Bouguer and Lambert and those of Beer and Bernard is called *Beer's law:* transmittance is the ratio of light intensity leaving the cuvette (l) to that incident on the cuvette (l_o).

$$T = \frac{l}{l_o}$$

However, since absorbance (A) is exponentially related to the reciprocal of transmittance,

$$A = \log \frac{1}{T} = \log \frac{l_o}{l}$$

Now the entire law, as it is used in absorption spectroscopy, can be summarized. The molecular constant of absorption, or the fraction absorbed at a given wavelength, is called absorptivity (A_s) and is a constant characteristic of the absorbing material. The thickness of the absorbing material or the length of the light path is expressed as b, and the concentration is expressed as c. The total absorbance is related to all these and is expressed as $A = A_s bc$, where concentration is expressed in grams per liter, light path in cm, and absorptivity in liters per grams times centimeters. Absorptivity can also be replaced by molar absorptivity (ϵ, expressed in liters per mole times centimeters), which is the constant corresponding to the absorbance of 1 mole of the absorbing material in solution, with a light path of 1 cm and a specific wavelength; thus, $A = \epsilon bc$.

A plot of absorbance versus concentration will be linear in the range for which Beer's law holds. When percent transmittance (% T) is plotted versus concentration, the plot is parabolic. Linearity of the decrease in % T with increase in concentration is achieved by a semi-log plot, or a plot of a log % T versus concentration (Figure 1–1).

Necessary Conditions for Beer's Law

The fundamental assumption of Beer's law is that each absorbing species or molecule absorbs monochromatic light independently of other molecules. Such a relationship can only hold in dilute solutions less than 0.02 M. In concentrated solutions, the

Figure 1–1

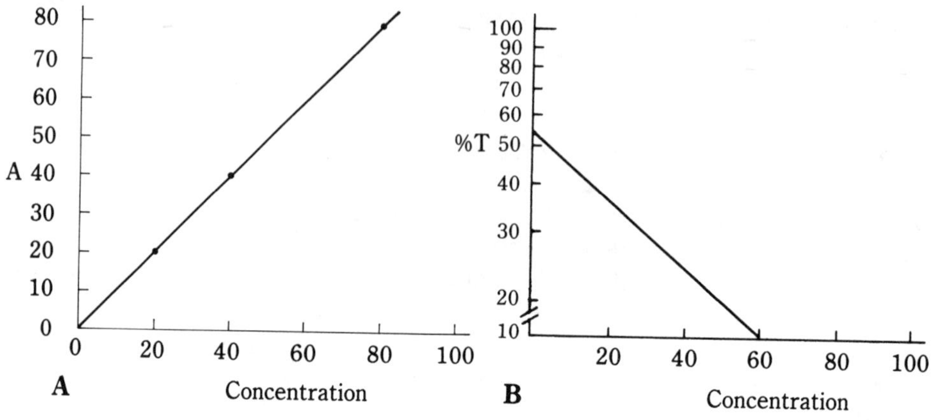

Beer's law. A: Linear plot of absorbance versus concentration. B: Semi-log plot of percent transmittance versus concentration. A = absorbance; %T = percent transmittance.

molecules collide with each other and dissipate some of the energy they have absorbed. In effect, in concentrated solutions no molecule absorbs light independently of another, and thus a plot of A versus concentration becomes nonlinear. Beer's law also postulates that there is no association or dissociation of the absorbing species and that equilibrium is essentially maintained during measurement. Hence, pH and temperature must be maintained during measurement to ensure that Beer's law is obeyed. Changes in refractive index during measurement and light scatter due to a turbid solution are two of the many factors that can cause deviations from Beer's law. Instrumental limitations in isolating a narrow wavelength band for absorption measurements (such as a large band pass or wider slit width), stray light (random reflections coming from various parts of the instrument and reaching the exit slit and detector), line voltage fluctuations, unstable light source, and nonlinear detector response all cause deviations from Beer's law.

Design of Photometers and Spectrophotometers

The components of a spectrophotometer include: light source; slits and associated optics; dispersing device for isolating the wavelength of light needed for measurement; cuvette containing the absorbing species and cuvette holder; exit slit; detector; and meter read-out, print-out, or recorder.

Spectrophotometers generally use either a grating or prism as a dispersing device; if the dispersing device is a filter, the instrument is referred to as a *filter photometer.*

Spectrophotometers or photometers with only a single alignment of the optical path from light source to detector are called *single-beam* instruments. A single-

Figure 1–2

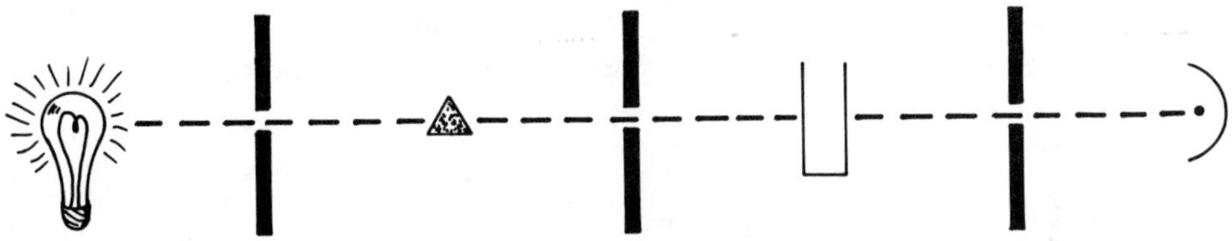

Single-beam spectrophotometer. The term *monochromator* refers not only to the dispersing device (prism, grating, or filter) but also to the associated optics and slits.

beam instrument is illustrated schematically in Figure 1–2. A single-beam instrument cannot simultaneously compare the absorbance of both the sample and the reagent blank, as is possible in double-beam instruments, and thus does not permit compensation for changes in light source intensity, line voltage fluctuations, and other instrumental drifts. Instead, there is a time lag associated between the sequential measurement of absorbance of the reagent blank and the sample.

Double-beam spectrophotometers have either a "true" double-beam-in-space design or a double-beam-in-time arrangement. Double-beam-in-space spectrophotometers have dual beams, for sample and reference, and dual detectors that compute the ratio of absorbance of the two beams (Figure 1–3). In this arrangement, the wavelength isolated by the dispersing device monochromator is split by beam splitters and mirrors and is directed toward the sample and reference beams and reaches the respective detectors; the signals of the two detectors are then compared.

A double-beam-in-time arrangement uses a single detector. A rotating chopper or mirror alternately focuses the sample and reference beams on the detector (Figure 1–4).

Both double-beam spectrophotometers are amenable to continuous recording of absorbance with time (kinetic measurements) or spectra over a range of wavelengths. The absorbance of the blank is automatically subtracted.

Rapid advances are being made in spectrophotometer design, including truly monochromatic light beams by means of tunable lasers, ultrafast scanning spectrophotometers, and extremely sensitive photodiodes. The optimum range of absorbance measurements in the first spectrophotometers was between 0.17 and 0.7 absorbance, whereas current detectors can resolve a 0 to 2 absorbance signal to 10^{-4} or 10^{-5} absorbance per count.

Spectrophotometer design can also permit recording of the first and second derivative of absorbance at designated wavelengths.[1] The potential of derivative spectroscopy has been exploited in the measurement of hemoglobin derivatives and,

Figure 1–3

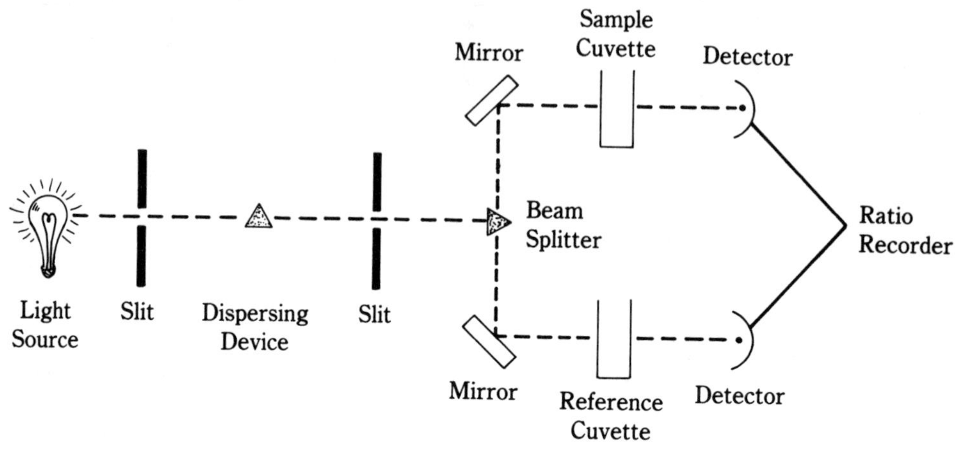

Double-beam-in-space spectrophotometer.

Figure 1–4

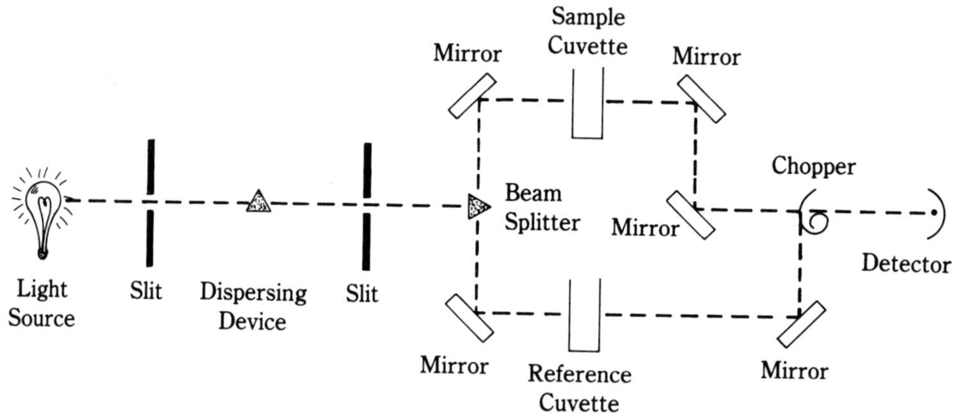

Double-beam-in-time spectrophotometer.

in general, in the elimination of baseline shifts, the reduction of blank absorbance, and turbidity. The first derivative is a measure of the rate of change of absorbance versus wavelength. Essentially this represents a plot of the slope of the spectrum where absorbance per nanometer of wavelength versus wavelength is plotted. The second derivative is essentially a derivative of the first derivative. Thus the second derivative spectrum is a plot of absorbance per nanometer squared versus wavelength. The second derivative spectrum is a measure of the curvature of the spectrum. It can effectively eliminate baseline tilts that may occur in the conventional plot of absorbance versus wavelength. "Zero-crossing point" first derivative spectrum is useful in the measurement of only one of the desired components in a two-component mixture.[2] Thus, methemoglobin can be measured independently of oxyhemoglobin at 414 nm. At this wavelength the first derivative value of oxyhemoglobin is zero irrespective of its concentration. As such, the first derivative value of oxyhemoglobin-methemoglobin mixture at the wavelength corresponding to the oxyhemoglobin "zero-crossing point" is directly related only to the methemoglobin concentration.

The measurement of hemoglobin fractions (hemoglobin, oxyhemoglobin, caroboxyhemoglobin, methemoglobin, sulfhemoglobin) can be conveniently performed with diode array spectrophotometers. These instruments can measure absorbances at five different wavelengths to account for the absorbances of the various hemoglobin fractions, plus an additional wavelength to correct for turbidity.[3]

References

1. O'Haver TC: Potential clinical applications of derivative and wavelength modulation spectrometry. Clin Chem 25:1548, 1979.
2. Taulier A, Levillain P, Lemonnier A: Determining methemoglobin in blood by zero-crossing point first-derivative spectrophotometry. Clin Chem 33: 1767, 1987.
3. Zijlstra WG, Buursma A, Zwart A: Performance of an automated six-wavelength photometer (radiometer OSM3) for routine measurement of hemoglobin derivatives. Clin Chem 34: 149, 1988.

2
Spectrophotometry

The main components of a spectrophotometer include a light source, a dispersing device, and a detector. A variety of light sources, varying in intensity, are available. The dispersing device can be a filter, grating, or prism. Many detectors are available, ranging from a barrier layer cell to a silicon photodiode.

Components

Light Sources

Tungsten lamps serve as incandescent light sources for the visible region of the spectrum.[1] At the usual operating temperature of such lamps (about 3000°K), however, about 85% of the energy emitted falls in the infrared region of the spectrum. The incorporation of low-pressure halogen vapor, such as iodine (as in tungsten-iodide lamps) or bromine, and the presence of a lamp envelope of fused silica increase the lifetime of the tungsten filament used in the currently popular quartz-halogen lamps.

Low-pressure, direct-current hydrogen or deuterium discharge lamps are used for the ultraviolet region. Such sources can operate at 165 nm with a fused silica optical window. The upper limit of the operating wavelength range for hydrogen or deuterium lamps is 360 nm, beyond which the hydrogen emission lines overlap the UV continuum emission. Deuterium discharge lamps have three to five times greater spectral intensity than the hydrogen lamp and have a longer lifetime than the latter. Voltage applied to such lamps excites the hydrogen or deuterium molecules which, in turn, emit their acquired energy in the UV region. Mercury or xenon arc lamps serve as sources in both the UV and the visible regions. The heat generated by such high-intensity sources, however, requires the use of a cooling or insulation device.

Figure 2–1

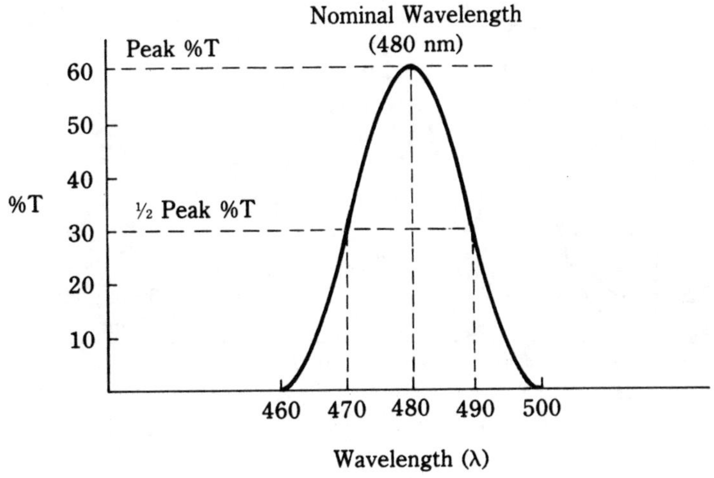

Band pass isolated by a filter. %T = percent transmittance.

Lamps age with use, and it is advisable to replace lamps routinely after a certain number of hours of operation. The vaporization of the tungsten filament can render the glass envelope of the lamp dark, necessitating replacement. In replacing lamps, care must be taken not to leave fingerprints. Lens paper can be used to hold the lamp and to wipe the lamp surface when the lamp is in position. Wherever the lamp and base socket design does not permit ready alignment, it is important that lamps be aligned manually.

Relatively new to spectrophotometry is the use of tunable laser beams as a light source; these sources are capable of providing true monochromatic light, thus allowing measurements at a discrete wavelength. Most light sources, however, are less specific, and their beams must be modified before they are useful in spectrophotometry.

Dispersing Device (Monochromator)

To isolate the desired wavelength, a filter, grating, or prism is used. The dispersing device with its associated optics constitutes the *monochromator.* The efficiency of a dispersing device is judged by the band pass it can isolate. The band pass is the range of wavelength isolated by the dispersing device at one-half of the peak transmittance measured at the nominal wavelength (Figure 2–1). In Figure 2–1, the peak transmittance achieved by the dispersing device (a filter) is 60% at the nominal wavelength (480 nm in this example). The range of wavelengths isolated at half peak transmittance (30% T) is between 470 and 490 nm, and the dispersing device in this case thus has an effective band pass of 20 nm. Note that the greater the peak transmittance, the narrower the band pass.

Figure 2–2

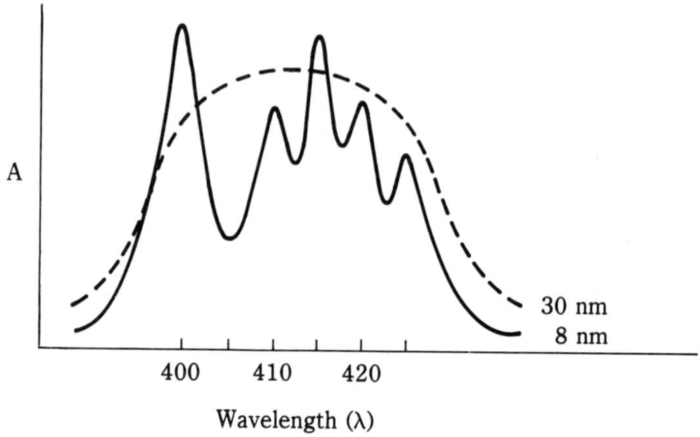

Effect of band pass on spectral resolution. A = absorbance.

The width of the band pass is critical in measuring sharp spectral peaks. Not only is the absolute absorbance of such spectral peaks underestimated by instruments with a wide band pass, but resolution of some narrow spectral peaks may be missed altogether. Figure 2–2 illustrates the spectra of a compound isolated with narrow and wide band pass spectrophotometers. The spectral peaks at wavelengths of 400, 410, 415, 420 and 425 nm are not resolved by the 30 nm band pass instrument; instead there is a broad peak in the 410 to 425 nm region. The underestimation of the absolute absorbance (A) of the peak at 400 nm by the 30 nm band pass instrument is due to the fact that the apparent absorbance measured is an average of the absorbance measured in the wavelength range that is isolated by the wide band pass instrument. In contrast, the 8 nm band pass instrument is not only able to isolate 5 spectral peaks, but also yields a greater absolute absorbance at 400 nm.

Filters. Filters isolate the wavelength of light either by absorption or by interference. Absorption filters function by selective absorption of unwanted wavelengths. One may speak in terms of "cut-off" filters that block light emitted either in the short or long wavelength of the spectrum. Composite filters are constructed using both short- and long-wavelength cut-off filters. Thus in Figure 2–1, light transmitted in the 460 to 500 nm range can be achieved by using 2 cut-off filters, one below 460 nm and the other above 500 nm.

Absorption filters generally consist either of solutions (liquid filters) or of organic dyes dispersed in glass (tinted glass). Another absorption filter is the Wratten filter, which consists of a film of dye-impregnated gelatin sandwiched between two layers of glass. Absorption filters have limited usefulness because of the wide band pass (35 to 50 nm) achieved by such dispersing devices and because of changes in the characteristics of dyes or gelatin with time.

Interference filters resolve the desired range of wavelengths of light by a process of selective reflection of unwanted wavelengths. Such filters generally consist of two pieces of glass that are coated on one side with a semitransparent silver film. A layer of dielectric, such as magnesium fluoride, is sandwiched between the two silver films. The thickness of the dielectric material (eg, magnesium fluoride) and its refractive index determine the range of wavelengths transmitted. A 200 nm dielectric layer of refractive index 1.4 will have a first order wavelength maximum at 560 nm as determined by the equation,

$$\lambda = 2nb/m,$$

where λ is wavelength, n the refractive index, b the thickness of the dielectric, and m the order number.

The wavelength transmitted by the interference filter and its harmonically related wavelengths are also transmitted. Thus, a first order filter transmitting a wavelength maximum of 600 nm will also transmit second and third order wavelength maxima at 300 nm and 200 nm, respectively.

The two layers of glass constituting the interference filter also serve as cut-off filters to eliminate overlapping orders of wavelength that would otherwise be transmitted by the filter. Interference filters transmit narrower bandwidths (10 to 15 nm) and have peak transmittance of 40% to 60%. Such filters are used widely in continuous-flow analyzers.

The band pass of an interference filter can be narrowed further (1 to 8 nm) by incorporating multiple layers of dielectric material with alternately low and high refractive indexes, producing a multilayer interference filter that can achieve peak transmittances from 60% to 95%. An interference filter can also be designed to select different wavelengths by depositing a wedge-shaped layer of dielectric material between the silver surfaces (continuously variable transmission interference filter). Moving the wedge permits the isolation of the desired wavelength. Unlike absorption filters, interference filters remove unwanted radiation by transmission and reflection and thus can be used with high intensity light sources.

Gratings. A grating consists of a large number of parallel lines or grooves that are closely spaced (600 to 1800 lines/mm) and ruled through diamond edged glass (transmission grating) or on a highly polished metallic surface, such as aluminum (reflection grating). Unlike transmission gratings, which transmit light, reflection gratings function like a mirror. Reflection gratings are inexpensive since replicas can be prepared from the original mold. Reflection gratings are widely used as dispersing devices in spectrophotometers.

The resolving power of a grating depends on the number of lines or grooves ruled on the grating surface. Light incident on each line or groove is spread out or diffracted. Each line acts as a prism and produces a spectrum. The angle at which light is diffracted depends upon the angle of incidence and the groove (the blaze angle). The smaller the distance between the adjacent lines or grooves, the more widely spread is the spectrum produced by constructive interference or reinforcement of wave fronts. The spectrum diffracted results from the incision of sections of wave fronts. To resolve light of different wavelengths, the grating is rotated to change the angle of incidence. Because gratings have a constant dis-

Figure 2–3

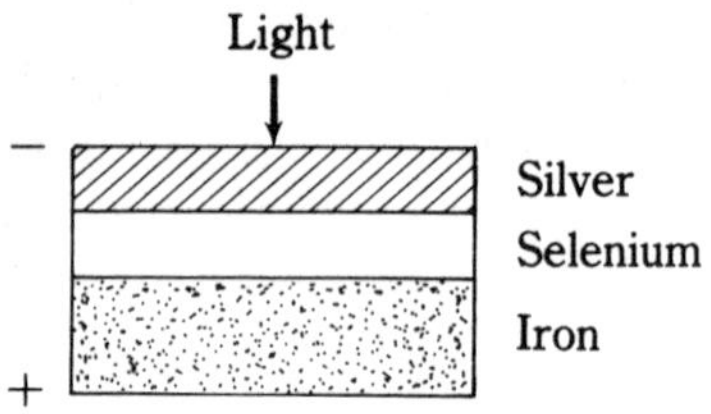

Barrier layer (photovoltaic) cell.

persion, the bandwidth or band pass will be constant throughout the spectrum. Gratings with an extremely narrow band pass can be produced by increasing the number of lines or grooves over their surfaces.

Prisms. A prism serves as a dispersing device by the variation of the index of refraction with wavelength. Shorter wavelengths, such as violet light, are bent more sharply than longer wavelengths, resulting in the crowding of longer wavelengths.

Unlike gratings, prisms produce a nonlinear wavelength scale. Because the band pass in a prism varies with wavelength, the slit width has to be adjusted to obtain a constant band pass. In recording spectrophotometers, the slit width is continuously narrowed as the spectrum is scanned from UV to the end of the visible region to ensure a constant band pass throughout the spectrum. Glass prisms are used in the visible region. To resolve UV spectra, prisms made of quartz or fused silica are used. The resolving power of a prism is determined by the base width of a prism and the dispersion power of the prism. In general, increasing the base width of the prism increases the resolution since the rate of change of refractive index with wavelength is nearly constant for a given prism. Gratings with extremely narrow band pass widths have greater resolving power than prisms. The base width of a prism would have to be extremely large and thus unwieldy to achieve the narrow band pass widths obtainable with a densely ruled grating.

Detectors

Four types of detectors are generally used: the barrier layer cell, the phototube, the photomultiplier tube, and the silicon photodiode.

Barrier layer cells. The barrier layer cell consists of a semiconductor layer, usually made of selenium, with an iron backing, that serves as the second electrode (Figure 2–3). A silver layer deposited on the selenium serves as the collector electrode. Light striking the selenium surface generates electrons that are collected by the silver layer and flow in the direction of the electron-deficient iron backing. The current produced is proportional to the radiant energy striking the semiconductor surface.

No external power supply is necessary for the functioning of the barrier layer cell. Since the current cannot be amplified, these types of cells are limited for use in single beam filter photometers, which have a fairly high level of illumination. Also, the barrier layer cell is fatigued with continuous use; this can be minimized by letting light strike the cell only when readings are taken by means of a controllable shutter. These cells also require time to equilibrate to operating temperature (warm-up), and they cannot respond to sudden changes in light intensity as occurs when a detector is alternately looking at reference and sample beams in a double-beam-in-time spectrophotometer.

Phototubes. A phototube comprises a metallic photocathode that is enclosed in a vacuum to prevent scattering of the electrons generated. The composition of the photocathode determines the spectral response; blue sensitive (350 to 650 nm) and red sensitive (600 to 950 nm) phototubes differ in the composition of the photocathode surface. A typical photocathode surface is silver oxide, cesium oxide, and silver metal. Light that strikes the photocathode surface generates electrons, since metals composing the surface are of low ionization energy and part easily with electrons. These electrons are collected and amplified by the external circuit. It must be remembered that both the lamp response and the phototube response will be different at different wavelengths. One therefore must readjust the 100% T reading with the blank whenever one changes the wavelength for measurement.

Photomultiplier tubes. A photomultiplier tube is a device with many cathode surfaces in series that permits a severalfold amplification of the initial light intensity incident on the tube's first surface (dynode).[2] Electrons produced at the first dynode are accelerated to the second dynode, and more electrons are produced. These in turn are amplified at succeeding dynodes, of which there may be as many as 9 to 16, before being collected and measured. The dynodes are coated with a metallic surface such as cesium-antimony or beryllium-copper. The acceleration of electrons to the successive dynodes is accomplished by applying voltage greater than that applied at the previous stage. A photomultiplier tube with a nine-stage dynode can result in a millionfold amplification of the original signal. These tubes have rapid response times and are ideal for use with double-beam spectrophotometers. Because of its powerful amplification effects, this type of tube emits a "dark current" even when the light source is turned off. Dark currents should be compensated for when readings are taken. These tubes should be shielded from bright illumination to prevent them from burning out.

Silicon photodiodes. A silicon photodiode detector consists of a high resistivity intrinsic silicon material sandwiched between two layers of silicon, called P and N, constituting respectively the top and bottom surface layers.[2] Thus this type of detector with an intrinsic silicon (i) sandwiched between P and N junction is referred to as a PIN silicon photodiode. The P layer has a protective coating of silicon dioxide (SiO_2) on its upper surface. Electrical contact is made through two metallic electrodes. The first is situated at the top and made out of a metal such as aluminum, and the other is at the bottom and usually constructed with gold. Figure 2–4 is a sketch of the design features of a silicon photodiode. In this type of detector, current

Figure 2–4

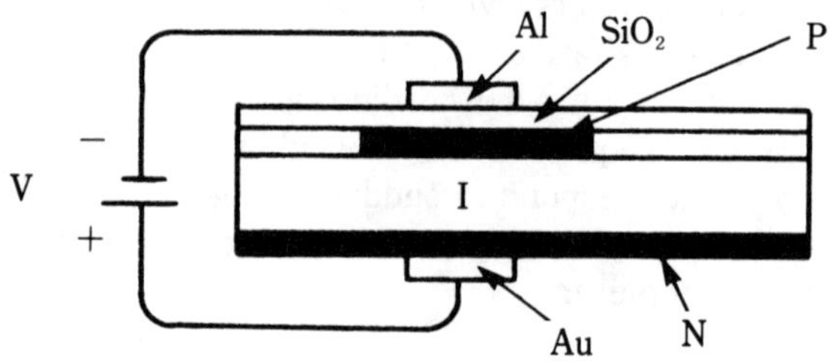

Schematic representation of a silicon photodiode. Al = aluminum electrode; SiO_2 = silicon dioxide protective layer; P = P-type silicon layer; I = intrinsic silicon material; N = N-type silicon layer; Au = gold electrode; V = external voltage circuit.

flow is initiated only when radiant energy strikes the P-type silicon layer. Electrons from the aluminum electrode and the external voltage source flow through the intrinsic silicon and N-type silicon layer to the gold electrode, thus completing the circuit. Current intensity is proportional to the number of photons reaching the intrinsic silicon material. Although the response of the silicon photodiode is restricted to the visible and infrared region of the electromagnetic spectrum, it is still an order of magnitude better than the response achieved with vacuum phototubes.

Diode array spectrometers theoretically have the ability to monitor the entire spectrum emanating from the cuvette. The spectrum is subsequently resolved by a grating into multiple wavelengths with each wavelength monitored by a diode that is part of a linear diode array. Diode array also permits bichromatic, trichromatic, and even polychromatic analysis to enable elimination of interference at adjacent wavelengths. Thus interferences due to hemoglobin (hemolysis) and lipemia can be handled by use of diode array. Diode array detectors have been used in high performance liquid chromatography (HPLC) for monitoring the spectra of urinary metabolites of interest in the study of inborn errors of metabolism.

With the development of spectrophotometers capable of scan times as precise as 5 μsec, highly sensitive detectors are available. For example, the Vidicon camera tube consists of an array of silicon photodiodes spaced approximately 15 mm apart. The current produced is directly proportional to the number of photons striking the array of photodiodes.

By using a truly monochromatic light beam and extremely narrow slit widths, it is possible to adapt photomultiplier tubes to photon counting. This technique yields improved precision and a higher signal-to-noise ratio.

Finally, the analog signal from the detector can be converted to a digital readout by using an analog-to-digital converter. The efficiency of such a converter depends upon the number of parts to which the absorbance signal is resolved. A 12-bit analog-to-digital converter can convert a 0 to 2 absorbance signal to a reading on the order of 10^{-4} absorbance per count.

Quality Control

Quality control of spectrophotometer performance is achieved by monitoring the accuracy of the wavelength selected for spectrophotometric measurements. In addition to wavelength accuracy, the suitability of cuvettes used for measurements, the response of the detector, and the presence, if any, of stray light should be evaluated. Fluctuations in meter readout and line voltage fluctuations should also be corrected.

Wavelength Accuracy

Monochromator alignment (wavelength accuracy) can be verified by a measurement of absolute absorbance. This can be accomplished by using filters or solutions that have absorbance peaks at specific wavelengths or by using light sources with sharp emission lines.[3]

Filters made of didymium glass or holmium oxide glass are widely used. Didymium filters have characteristic wavelengths of absorption in the visible region (573, 586, and 685 nm), whereas holmium oxide filters have absorbance peaks in both the UV and the visible region. Several solutions are available for checking absolute absorbance. Among these are alkaline potassium chromate (40 mg/liter), which has absorbance peaks at 340, 375, 400, and 450 nm, and acid solutions of cobalt chloride or nickel sulfate, both of which have an absorbance peak at 510 nm.

Wavelength accuracy of narrow band pass instruments are best checked by using emission sources with sharp spectral lines.

Wavelength accuracy of spectrophotometers that utilize a deuterium lamp as a UV emission source can be easily checked because such lamps have two sharp spectral lines in the visible region (486 and 656 nm) and a less intense line at 380 nm.

Replacing the light source with a quartz-mercury arc lamp, which has sharp spectral lines between 205 and 1014 nm, theoretically permits the best check of wavelength accuracy, but in practice it is cumbersome to use with most precision double beam spectrophotometers.

Wavelength accuracy of spectrophotometers that use prisms as dispersion devices should be checked at a minimum of three wavelengths, since wavelength dispersion is nonlinear. In contrast, gratings have linear dispersion throughout the spectrum and need to be calibrated at only two wavelengths. A rigid check of wavelength accuracy is mandatory after an instrument has been moved or after replacement of its light source. Routine daily checks can be accomplished satisfactorily with one of the glass filters or solutions (sealed in cuvettes) that are available as standard reference materials from the National Bureau of Standards.

Cuvettes

For measurements in the visible region, glass cuvettes are adequate. The use of a specialty glass can extend the starting wavelength down to 300 nm. Quartz cuvettes extend the starting wavelength down to 210 nm, and fused silica down to 185 nm.

Figure 2–5

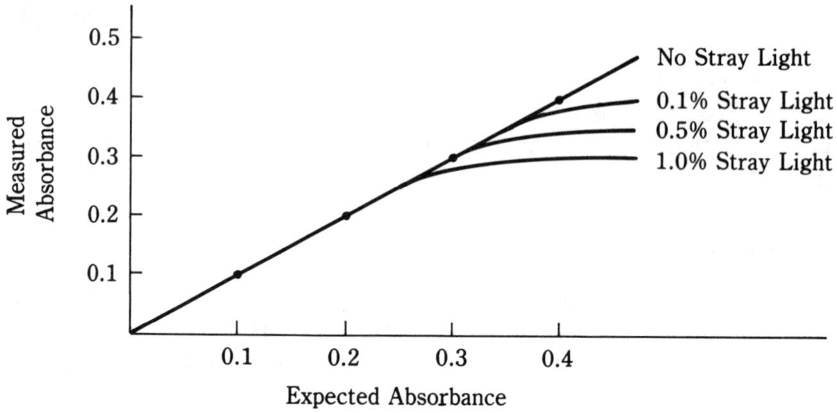

Effect of stray light on detector response.

Care should be taken in the handling of cuvettes; fingerprints should not be left on the optical surfaces. Cuvettes should be matched with stable colored solutions. Matched cuvettes that do not agree within 0.01A of each other have scratched surfaces and should be discarded. Round cuvettes should be marked and always introduced in the same position to compensate for fluctuations in front to back rotation. Errors associated with the cuvette holder, such as loose attachment, should be corrected.

Detector Response

Quality control of detector response is performed with linearity checks using solutions of different concentrations. For instance, copper sulfate solutions of 1%, 2%, and 3% should theoretically give an absorbance ratio of 1:2:3 at 650 nm if detector response is linear. Some of the solutions recommended earlier for measuring wavelength accuracy can also be used at three or four concentration levels to monitor detector response. Any one of the green food coloring dyes that have absorbance maxima in the UV region and two in the visible region are ideal for measurement of detector response.

Stray Light

Even when the detector response is optimal, however, nonlinearity can be observed when measurements are made with calibrating solutions. This nonlinearity is caused by stray light randomly reflected onto the detector from various parts of the instrument. Stray light can also be caused by dust, leakage of extraneous light, or overlapping orders of wavelength transmitted by a grating. Figure 2–5 illustrates the effect of stray light on detector response.

A solution or filter used to measure stray light should be opaque at or below the wavelength at which stray light is measured. Ideally, if no stray light reaches the detector at the wavelength used for absorption measurement, no light will be transmitted. The percentage of light that is transmitted under these conditions would directly correspond to the stray light reaching the detector.

A saturated solution of lithium carbonate (opaque below 225 nm) or 1% sodium iodide solution (opaque below 240 nm) can be used to measure stray light in the UV wavelength range. Sharp cut-off silica filters are available to measure stray light in both the UV and visible region. Stray light can be minimized by removing dust from the surfaces of the instrument, preferably with a camel hair brush and tubing through which gentle vacuum can be applied.

Meter Readout

Some spectrophotometers have a meter readout. Erratic deflection of the meter needle indicates that moisture has collected on the resistors. This can be corrected by placing an indicating silica gel pouch in the phototube compartment. The silica gel can be regenerated when its capacity to absorb moisture is exhausted.

Electric Current Fluctuations

Finally, fluctuations in electrical current can occur that can be compensated by using a voltage stabilizer. Some instruments have balance circuits that compensate for this.

In addition to quality control considerations relating to spectrophotometer components—from the light source to the detector—other measures that can reduce the cumulative errors that can result include the following:

- Maintenance of temperature control in enzyme assays
- Assurance of the integrity of the sample itself (ie, that it is free of air bubbles and turbidity)
- Control of solution concentration to within a range where Beer's law is followed.

References

1. Bauman RP: Absorption Spectroscopy. New York, Wiley, 1967.
2. Willard HH, Meritt LL, Jr, Dean JA, Settle FA, Jr: Instrumental Methods of Analysis, ed. 6. New York, Van Nostrand, 1981.
3. Edisbury JR: Practical Hints on Absorption Spectrophotometry. New York, Plenum Press, 1967.

3
Fluorometry

Origin of Fluorescence

Molecules at rest are in the ground state. In this state two electrons in the molecular orbit have opposing spins, essentially canceling each other and thereby existing in the singlet state (S_0). Energy supplied in the form of high energy photons from a suitable light source elevates the singlet state electrons to higher energy levels, called excited singlet states (S_1, S_2), within 10^{-15} seconds. The excited electrons rapidly dissipate their excess energy through collisions with other molecules (Figure 3–1) and reach the minimal vibrational energy level of the lowest excited state (S_1) by mechanisms such as the loss of excess vibrational energy as heat by vibrational relaxation. The energy remaining with the electrons in the lowest excited state is emitted as they return to the ground state. This luminescence phenomenon, occurring within 10^{-9} to 10^{-5} seconds, is referred to as *fluorescence.* A luminescence phenomenon lasting longer than 10^{-4} seconds is called phosphorescence and involves, unlike fluorescence, a change in the spin of the electron (intersystem crossing) as the excited singlet state electron is transformed to the triplet state prior to returning to the ground state. Fluorescence emission has a longer wavelength (lower energy) since some of the energy acquired by the excited molecule is dissipated in the form of excess vibrational energy before fluorescence emission occurs. This increase in wavelength of fluorescence emission, as compared to the excited wavelength, is referred to as Stokes' shift.

Excitation and fluorescent emission spectra have a mirror-image relationship to each other. The intensity of fluorescence emission is directly proportional to the strength of the incident excitation source and the quantum efficiency of fluorescence (ϕ_F). The latter is a ratio of photons that fluoresce to the photons absorbed when the molecules are excited. Quantum efficiency thus represents the fraction of excited

Figure 3–1

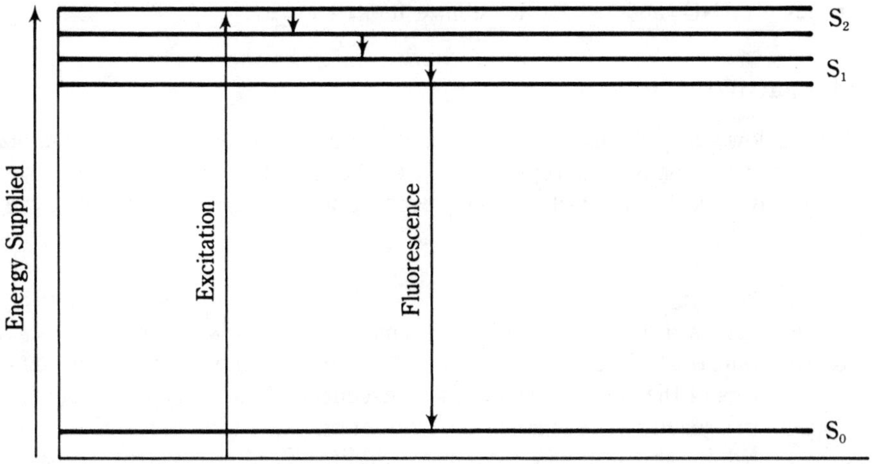

Summary of the events leading to fluorescence emission. The small arrows at top depict the dissipation of energy as excited electrons reach the minimum vibrational level of the lowest excited state.

molecules that undergo fluorescence emission, as opposed to other forms of energy transformation, eg, collision deactivation and phosphorescence via intersystem crossing. Each fluorescent species has its own quantum efficiency that is constant. Molecules emitting most of their absorbed energy as fluorescence will have quantum efficiency values close to unity, and will be efficient fluorophores.[1]

Factors Affecting Fluorescence

A variety of factors affect the ability of a molecule to fluoresce, including the molecule's structure. The temperature during measurement and the pH of the sample influence emission, and the presence of impurities can affect the fluorescence signal. The concentration and viscosity of sample and the solvent used to dissolve the sample are among the other variables that affect fluorescence.

Structure

Certain structural features must be present in a molecule for it to fluoresce. In general, molecules stabilized by resonance tend to fluoresce. Thus, those aromatic compounds for which many resonance stabilized structures are possible, and compounds with multiple alternating double bonds, can be expected to fluoresce. Rigidity in a molecule favors fluorescence by reducing internal vibrations; rigid metal chelates (eg, calcium-calcein-EDTA complex) fluoresce. Highly fluorescent compounds such as fluorescein have rigid structures. Electron donating substituent groups in a molecule such as OH and NH_2 promote fluorescence, while electron withdrawing groups such as NO_2 and COOH destabilize fluorescence.

Temperature

Because at low temperatures the collisional frequency of molecules is minimized, fluorescence emission of a fluorophore will be enhanced. Increasing the temperature will obviously have the opposite effect, resulting in a reduced fluorescence yield.

pH

A change in pH will affect the molecular charge. If the change in pH results in greater resonance stabilization (the ability to distribute the charge over multiple structural forms of the same molecule), fluorescence will be enhanced. Conversely, if change in pH results in resonance destabilization, then the ability of a molecule to fluoresce will be affected. The principle of utilization of fluorescent indicators in acid-base titrations depends on the effect of change in pH on the fluorescing property of the indicator. The indicator could be colorless at one pH and highly fluorescent at another.

Impurities

Certain metallic ions, such as Cu^{2+} and Ni^{2+} will not form fluorescent complexes because of their paramagnetic nature (ie, attracted to a magnetic field, due to the magnetic moment characteristic of an unpaired electron). Metals such as magnesium and zinc, however, can form highly fluorescent complexes (eg, zinc erythrocyte protoporphyrin). A paramagnetic species, such as oxygen, will, in solution, decrease or quench fluorescence. Interaction of fluorescence emitted by a molecule with an impurity can lead to quenching.

Concentration

Fluorescence intensity is linearly related to concentration in dilute solutions (less than 10^{-5}M). In concentrated solutions the emitted fluorescence is dissipated by the frequency of molecular collisions (self-quenching) and the fluorescence-concentration calibration curve becomes nonlinear. An additional reason for nonlinearity of the calibration curve in concentrated solutions is the occurrence of the phenomenon of self-absorption; the fluorescence emitted by the molecules is also absorbed by the molecules, especially where the wavelengths of absorbance and fluorescence are close to each other.

Viscosity

A viscous solvent, by promoting the rigidity of a fluorescing molecule, will increase the intensity of fluorescence. Thus, the fluorescence of a sample in glycerol will be greater than that observed in aqueous solution.

Figure 3–2

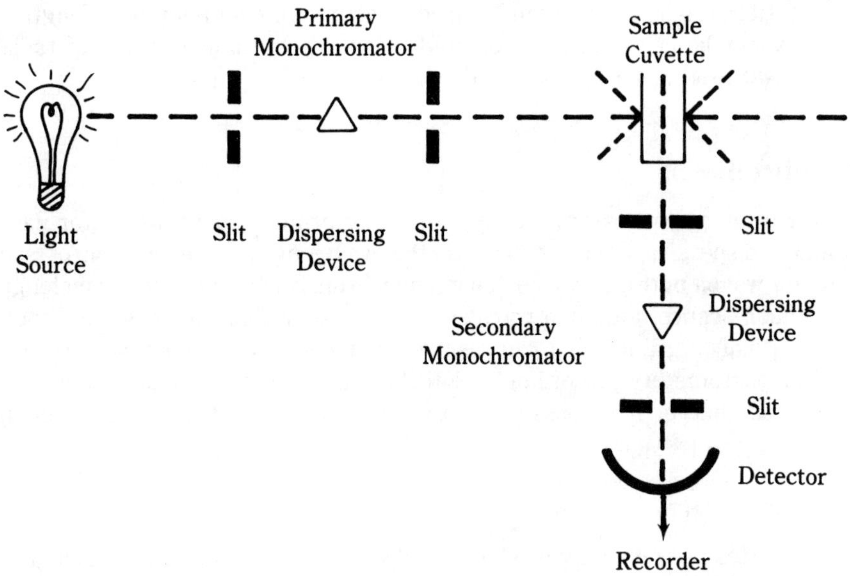

Design of a fluorometer.

Solvent

The wavelength at which a compound will fluoresce and its fluorescence intensity will be dictated by the nature of the solvent. Solvents whose electron clouds interact (Van der Waals forces) with the excited molecule will provide adequate time for dissipation of energy of molecular collisions, thereby decreasing fluorescence emission.

Design and Components

The basic design features of a fluorometer (Figure 3–2) are: a light source; slits to focus the light; a primary dispersing device, such as a filter, grating or prism to isolate the wavelength of light needed to excite the sample to fluoresce; the sample cuvette; slits to focus the emitted fluorescence; a secondary dispersing device, such as a filter, grating, or prism to isolate the wavelength of emitted fluorescence light; a detector; and a readout device, such as a meter or recorder.[2]

Light Source

Spectrofluorometers generally utilize an intense, stable, continuous beam emitted by a high-pressure xenon arc lamp providing the capability to choose an excitation wavelength in the ultraviolet (UV) or visible region.

Filter photometers generally use low-pressure mercury vapor lamps as a source. Mercury emits distinct line spectra at several discrete wavelengths in the UV and visible regions; these can be used as source of excitation wavelength. The mercury vapor lamp can also be adapted to emit a continuous beam of radiation between 300 to 405 nm by coating the lamp with a white phosphor.

Monochromator

Two sets of monochromators are required. The primary monochromator includes slits and a dispersive device to isolate the wavelength for excitation of sample (excitation wavelength); the secondary monochromator isolates the wavelength of emitted fluorescence (emission wavelength). In spectrofluorometers the dispersive device is usually a grating that can resolve a narrower band pass than can a prism.

Filter photometers use either an interference filter to isolate a required mercury emission line or band pass filters to resolve a selected range of wavelength.

Cuvettes

Quartz cuvettes are generally used, especially when the excitation wavelength chosen is in the UV region (glass absorbs UV). Optically clear plastic cuvettes may also be used provided the solutions used do not interact with the cuvette material. The cuvette compartment is usually painted a dark shade to minimize scattered radiation reaching the detector.

Detector

A phototube or a photomultiplier tube is used. Note that the detector is not aligned in the same path as the light source. Instead the detector is at right angles (90°) to the cuvette. Such an alignment is possible since fluorescence is emitted in all directions (360°). In the conventional arrangement in a spectrophotometer, the detector is in the path of the light source and cuvette. The advantage of the 90° detector geometry is that the primary excitation beam and light reflected from the cuvette surfaces (with the exception of the surface facing the detector) are excluded from reaching the detector.

The detector can also be aligned at an angle of 37° from the cuvette thus permitting measurement of front surface fluorescence of solids, turbid solutions, and nearly opaque materials, such as samples collected on filter paper. Instruments designed for measurement of front surface fluorescence have made it possible to measure fluorescing constituents (eg, zinc erythrocyte protoporphyrin) in drops of blood collected on filter paper.

Readout

A meter or recorder is used to plot the relative fluorescence emission. Calculations are based on standards of known concentration, and calibration curves represent a plot of relative fluorescence vs concentration. A typical calculation using a single standard is:

$$\frac{\text{fluorescence (sample} - \text{blank)}}{\text{fluorescence (standard} - \text{blank)}} \times \text{concentration of standard}$$
$$= \text{concentration of sample}$$

The meter readout can be adjusted to read the concentration of the sample directly after calibrating with a standard of known concentration.

As relative fluorescence is measured, the path length of the cell is not critical. However, the slit widths are critical because they regulate both the excitation intensity reaching the sample and the emission intensity reaching the detector.

Relative Merits

The fluorometer design depicted in Figure 3–2 does not allow for fluctuations of light source or variations in detector response. This deficiency can be overcome by measuring the ratio of sample to reference beams. In such "ratio fluorometers," light from the source is split into sample and reference beams, each reaching their respective detectors. The signals of the sample and reference photomultiplier tubes are compared and the ratio readout compensates for fluctuations during measurement. Another approach used in some filter fluorometers balances the sample fluorescence emission beam with a calibrated light path, with both beams "seen" alternately by a photomultiplier tube. A fluorescence readout dial is rotated as the calibrated beam is adjusted to equal the fluorescence emission beam and the photomultiplier response is nullified. The null balance approach effectively overcomes fluctuations in light source and detector response.

The instrument design depicted in Figure 3–2 has no provision for simultaneously comparing a reference sample to the sample being measured. Precision double-beam-in-time spectrofluorometers permit the detector to alternately compare reference and test samples, thus providing a differential readout.

Some spectrofluorometers have design features that minimize light scatter and permit use of wide slits, and yet provide good resolution. This is achieved by using dual gratings for resolving excitation and emission wavelength (double monochromation).

Spectrofluorometers have advantages over filter fluorometers, especially when it is necessary to establish first the excitation and fluorescence emission wavelength of a fluorescing sample.

Since a spectrofluorometer permits easy selection of wavelength, the maximum excitation wavelength can be established by keeping fixed the excitation wavelength at which fluorescence is observed and scanning the fluorescence emission spectrum. The wavelength at which maximum fluorescence was observed is kept fixed, and the excitation spectrum is scanned to determine the wavelength of maximum excitation.

The narrow band pass achieved with grating spectrofluorometers permits selection of both excitation and fluorescent emission wavelengths different from that of an interfering substance possessing a nearly overlapping excitation and emission spectrum.

Applications

Fluorometry can be used not only for the measurement of constituents that fluoresce but also for species that can be rendered fluorescent. Examples of naturally fluorescing substances include drugs such as quinidine, propranolol, salicylates, and barbiturates; porphyrins (zinc erythrocyte protoporphyrin as an index of lead poisoning); reduced coenzyme NADH and NADPH; and amino acids (eg, tyrosine and tryptophan).[3]

Among substances that can be rendered fluorescent by chemical reaction are cortisol, estrogens, and catecholamines. Metals such as calcium and magnesium can be determined fluorometrically after chelation. Reaction with fluorescent samples is the basis for the measurement of creatine and phenylalanine.

Immunoassays have been developed using fluorescence labels for the measurement of specific proteins (lgG, lgA, lgM, C_3, C_4), T_3 and T_4, drugs (phenobarbital, phenytoin), antibiotics (tobramycin and gentamicin), as well as antinuclear antibodies (ANA) and anti-DNA antibodies. These fluorescent immunoassays can be performed either without separation steps (homogeneous fluoroimmunoassays) or they may require washing and separation steps (heterogeneous fluoroimmunoassays). Fluorescence can be measured with a fluorometer either in solution or on solid polymeric surfaces that can be easily inserted into a fluorometer designed to receive them.

The availability of synthetic fluorogenic substrates with fluorophores, such as 5-amido isophthalic acid dimethylester or methyl umbelliferone that are cleaved by specific coagulation enzymes have revolutionized measurement concepts in coagulation.

Fluorescing species separated by paper or thin layer chromatography can be identified by examining the chromatogram under UV light. Fluorescence detectors have also been adapted to high performance liquid chromatography (HPLC) permitting automated analysis of several fluorescent species in a single sample. Fluorescence detection is also the basis of some instruments that separate and sort out various types of cells and cellular constituents.

Relative Merits of Fluorescence

Fluorescence measurements are characterized by high sensitivity permitting measurement of nanomole concentrations of fluorescing species. The introduction of fluoroimmunoassays has extended sensitivity down to the picogram range. Fluorescence emission is measured against an essentially negligible blank background (unlike spectrophotometry). Fluorescence can be measured in highly dilute fluorescing solutions without sacrificing sensitivity (unlike absorption measurements, in which detector sensitivity becomes a limiting factor in measuring small absorbance signals).

Fluorescence measurements have a high degree of specificity. They can be performed in the presence of interfering substances that have overlapping absorption spectra, or similar excitation spectra, as long as the fluorescence emission spectra of the species measured is at a wavelength different from those emitted by the interfering substances.

Fluorometric measurements can be standardized by using fluorescent compounds of known quantum yield, such as fluorescein, quinine sulfate, or rhodamine B. Problems in fluorometric measurement include the need to maintain the pH and control temperature during measurement, and the possibility of quenching and self-absorption in concentrated solutions. There is also the possibility that the intensity of the excitation source will decompose the compound whose fluorescence is to be measured. The precision of fluorometric measurements is inferior to that obtained with absorption measurements. These disadvantages, however, are outweighed by the sensitivity, specificity, and the wide variety of fluorescence assays available to the clinical laboratory.

References

1. Guilbault GG: Practical Fluorescence Theory. Methods and Technique. New York, Marcel Dekker, 1973.
2. Willard HH, Meritt LL, Jr, Dean JA, Settle FA: Instrumental Methods of Analysis, ed 6. New York, Van Nostrand, 1961.
3. Rubin M: Fluorometry and phosphorimetry in clinical chemistry. Adv Clin Chem 13:161–269, 1970.

4
Flame Photometry

The emission of neutral atoms in a flame forms the basis of flame photometric measurements. The ease of emission of an atom in a flame is related to its ionization potential, and alkali metals with low ionization potentials, such as sodium, potassium, lithium, and cesium, are best suited to this method of measurement. Even so, flame emission is a difficult process because the ratio of the number of atoms in the excited state to those in the ground state, although dependent on flame temperature, is infinitesimally small; at a flame temperature of 2000°K this ratio is approximately 4×10^{-4} for cesium and 1×10^{-7} for sodium.

Emission Wavelength for Analysis

The flame's heat excites orbiting electrons in an atom from the lower energy ground state to the excited state. Electrons can be excited to any one of the several energy levels of the excited state. If the electron is excited to the lowest energy level in the excited state, it can theoretically emit all of the energy it has acquired in the form of light as it returns to the ground state. Such emission lines will be intense. The 589/590 nm doublet of sodium, the 767/769 nm doublet of potassium, and the 671 nm emission line of lithium represent wavelengths corresponding to the electronic transition from the lowest excited state to the ground state and represent wavelengths of maximum emission. The longer wavelengths of these emissions are consistent with the small energy differences between the ground state and the lowest excited state.

Emission lines that represent a transition of the electron from the ground state to the higher excited states are not sensitive emission wavelengths for measurements, because a fraction of energy acquired by the electron is dissipated as the

Figure 4–1

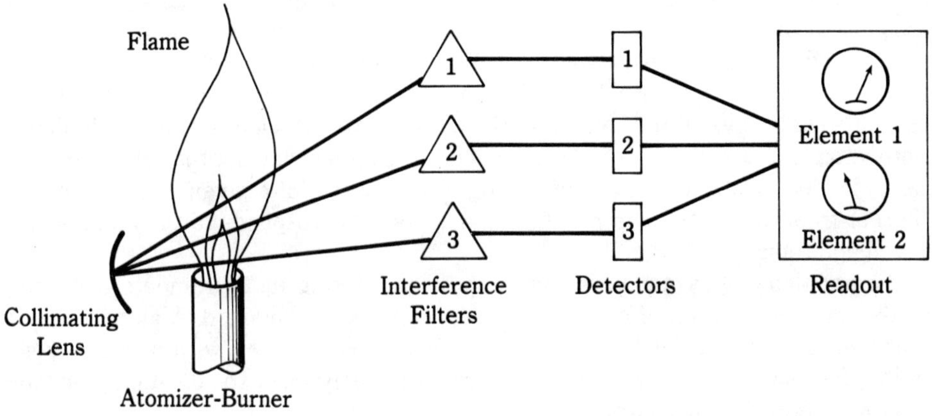

Basic design features of a flame photometer. Interference filter 2 isolates the emission of an internal standard. Filters 1 and 3 isolate the emission of two different elements, such as sodium and potassium or lithium. Not shown are cylinders for fuel gases.

electrons return from the higher excited state to the lowest excited state and finally to the ground state.

In addition to line spectra, which result from electronic transition of the atom, two other types of spectral emission, known as continuous and band spectra, are evident. Both contribute to the flame background. Incandescent solids in the flame give rise to spectra of similar intensity over a range of wavelengths and to continuous spectra. Metal oxides in the flame and gaseous molecules cause band spectra that are composites of vibrational, rotational, and electronic transitions of these molecules.

Atoms in the flame undergo electronic transitions rather than the rotational and vibrational modes of transition of molecules. Atomic emissions are not truly lines because of line broadening, which results from a variety of factors including time spent by the atoms in the excited state (natural broadening), the limitations of the detector in the face of random thermal motions of the atoms (Doppler broadening), and dissipation of energy by collision with other atoms in the flame (collisional broadening).[1]

Design and Components

The basic design features of a flame photometer include fuel gas cylinders and pressure regulators, compressed air, an atomizer-burner, a monochromator (interference filters or grating), detectors (photocells, photomultiplier tubes), and a concentration readout device (Figure 4–1).

Gases

Low-temperature flames, usually a mixture of propane and air reaching temperatures of approximately 1925°C, are used in the flame photometric analysis of alkali metals. Higher temperatures would result in the ionization of a fraction of neutral atoms that would otherwise be excited and emit radiation characteristic of the element to be analyzed, thus affecting the sensitivity of the measurement. In flame photometry, the wavelength of emission of neutral atoms, and not that of the ionized atoms, is measured. In fact, the wavelength of emission of a given ionized atom is different from that of the neutral atom and actually corresponds to the atom of the preceding atomic number.

Fluctuations in the temperature of the flame during measurements will alter the degree of excitation of atoms and affect the results obtained. Maintenance of a constant and proper ratio of fuel to oxidant (air) is achieved with pressure regulators. Two-stage pressure regulators are commonly used to maintain constant gas pressure during measurements.

Atomizer-Burner

The most efficient method of atomization is the reduction of the sample to an aerosol, creating an atomized mist or fog that enters the flame. This is achieved by aspirating the sample through a capillary under reduced pressure generated by a flowing gas stream. The aspirated sample droplets are further broken down in the spray or atomizing chamber by the impact resulting from the collision of the particles against the sides of the chamber. In this manner, large droplets (larger than 20 μm) are discarded into the waste receptacle, and only an aerosol of atoms mixed with the fuel gas-air mixture reaches the flame. Efficient atomization is achieved in this type of atomizer-burner combination at the expense of sensitivity because less than 10% of the original sample reaches the flame.

Indeed, sensitivity can be increased by aspirating all of the sample into the flame as is done with integral–atomizer-burners. In this type of arrangement, the sample, fuel gas, and air are all delivered separately to the flame area. In spite of the theoretically increased sensitivity, the configuration suffers from the noise that results from the introduction of large droplets into the flame. In addition, hotter flames result, decreasing by ionization the population of atoms available for emission. For this reason the integral–atomizer-burner is rarely used. The rate at which the sample is aspirated into the flame should be maintained throughout the measurement. The optimum flow rate is generally achieved by maintaining a 10 lb/in^2 pressure with compressed air.

Monochromator

In the flame photometers commonly used in the clinical laboratory, isolation of the wavelength of spectral emission is usually achieved with interference filters. For research purposes, flame spectrometers utilizing gratings allow improved resolution of spectral lines of emission. Monochromation in both filter and grating flame spectrometers is achieved by a combination of collimating lenses and mirrors.

Detector

Either barrier layer cells (photocells) or photomultiplier tubes are used as detectors. The barrier layer cell is adequate for the detection of levels of sodium and potassium found in serum and constitutes the detector component in many of the flame photometers routinely used in the clinical laboratory. The photomultiplier tube offers greater sensitivity because, unlike the barrier layer cell, it is capable of amplifying very small amounts of signal current severalfold and is not subject to fatigue. Photomultiplier tubes are used with research instruments offering greater sensitivity of measurements.

Readout

The meter readout is calibrated to display the emission signal of the sample in concentration units. Single-beam instruments (ie, having no provision for comparing the signals from both the sample and an internal standard) cannot measure two or more elements simultaneously. To measure the concentrations of multiple elements, the corresponding wavelengths must be readjusted and the sample reanalyzed. Furthermore, because of the lack of an internal standard, fluctuations in flame temperature and sample aspiration rate are not compensated for and thus introduce an error in the results.

Consequently, single-beam flame photometers are rarely used in the clinical laboratory. The readout devices in instruments that have provision for an internal standard measure the ratio of the emission signal of the test elements to that of the internal standard, thus simultaneously providing results in concentration units of two desired elements in the sample. By measuring the ratio, any fluctuations in flame temperature and sample aspiration rate are canceled out since they affect the element that is measured and the internal standard to the same extent.

Flame

Reactions

The sample that enters the flame as a mist of atoms is instantly vaporized, leaving behind a population of dry metal salts. These salts are then raised to a gaseous state and subsequently dissociate to give off neutral atoms. A fraction of the neutral atoms is excited by the thermal energy of the flame and emit the characteristic line spectrum of the element, the intensity of which is related to the concentration of the element in the sample. It is important that the neutral atom is excited to emission and not ionized, because ionization will reduce the fraction of neutral atoms available for emission and thus reduce the sensitivity of the measurement. Figure 4–2 summarizes the sequence of reactions that occurs in the flame with a salt in the sample.

Characteristics

There are four distinct regions in the flame (Figure 4–3): the preheating area, the primary combustion area (inner cone), the interconal area, and the secondary com-

Figure 4-2

$$NaCl\ (spray) \rightleftharpoons NaCl\ (gas) \rightleftharpoons Na^0\ Cl^0 \begin{cases} \nearrow Na^*\ \text{(excited atom)} \\ \searrow Na^+\ \text{(ionized atom—undesirable)} \end{cases}$$

Sequence of reactions in flame photometry.

Figure 4–3

Flame characteristics.

bustion area (outer mantle). Combustion of the fuel-air mixture entering from the burner is initiated in the preheating zone by the conduction and diffusion of radicals from the inner cone or the primary combustion area. The bright bluish green inner cone is populated mainly by flame gases such as carbon monoxide, carbon dioxide, hydrogen, nitrogen, water, and, to a lesser extent, by hydrogen and oxy and hydroxy radicals. Atomic emission is achieved in the interconal zone, where the flame is less luminous, and thermal equilibrium is maintained. The monochromator in the instrument is aligned in the path of the interconal zone so that it can isolate the spectral lines of elements issuing from this region. The secondary combustion zone, or outer mantle, is the result of entrapped air, which oxidizes flame gases such as carbon monoxide to carbon dioxide and completes the process of combustion. The blue-violet outer mantle is useless from the standpoint of atomic emission.

Interferences

Spectral

Molecules in the flame emit their own spectra (band spectra) and contribute to the flame background. Calcium, for instance, forms oxides and hydroxides that emit at wavelengths close to those of the spectral lines chosen for sodium and lithium measurements.

Such background spectral interference is in practice not quite so serious and can be corrected for electronically unless the concentration of interfering molecules is extremely high. In concentrated solutions, a fraction of emitted radiation is absorbed by some of the atoms that are present in excess, thus decreasing the intensity of emission reaching the detector and the sensitivity of the measurement. This phenomenon is called "self-absorption," which in practice is overcome by the commonly used 1:200 dilution of serum in the measurement of alkali metals.

Ionization

Ionization is undesirable because it reduces the fraction of neutral atoms available for emission and thus lowers the sensitivity of the measurement. Ionization interference can be overcome by adding an excess of an easily ionized element, such as cesium, to the assay mixture. Even at low concentrations, the excess of electrons thus generated will shift the equilibrium of the test elements in favor of neutral atoms and thus keep ionization to a minimum. In practice, at the relatively low temperature of the air-propane flame, ionization interference is not serious and does not require the addition of an ionization suppressant. The excess of internal standard used in such measurements sufficiently minimizes ionization interferences due to fluctuations in flame temperature.

Physical

Physical interferences, such as the increasing size of droplet formation, can have a cooling effect on the flame and thus reduce the emission intensity and sensitivity of the measurement. Fortunately, the commonly used atomization devices permit aspiration of minute droplets in the form of an atomized fog, thereby reducing the possibility of large droplets reaching the flame. Other physical interferences, such as increased viscosity and surface tension, affect aspiration rate. These, of course, are overcome by the 1:200 dilution that is commonly employed and, further still, by the incorporation of small amounts (0.02% to 0.04%) of a nonionic surfactant to reduce the surface tension and thus improve the efficiency of aspiration.

Chemical

Chemical interference caused by anions and cations needs to be considered. Anions such as chloride can reduce the emission intensity of sodium and potassium, but the interference is minimal at the dilution ratio commonly used for measurements. Interference caused by cations (the increase in the relative percentage of potassium atoms that emit in the flame due to a high concentration of sodium) can be overcome by incorporating in the sample a radiation buffer, which is usually the internal standard.

Internal Standard

Lithium is used as an internal standard in the analysis of sodium and potassium. For the analysis of lithium, potassium is used as an internal standard. Cesium could also serve as an internal standard for the analysis of sodium, potassium, and lithium.

The internal standard has a dual purpose. First, since the ratio of emission intensity of the test element to that of the internal standard is taken, any fluctuations during measurement in the temperature of the flame caused by changes in fuel and air pressures, changes in sample aspiration rate, or fluctuations in line voltage, affect the test element to the same extent as the internal standard and therefore cancel out.

Second, because the internal standard is present in excess relative to the test element, it also serves as a radiation buffer. Thus for example, when the concentration of sodium in the sample is high, emission energy caused by the high concentration of sodium is used first to excite some of the excess atoms of the internal standard, thereby preventing the transfer of energy to potassium atoms and influencing its emission. By serving as a radiation buffer, the internal standard ensures that the intensity of emission of potassium atoms in this example is related to its concentration and is not dependent on the concentration of sodium in the sample.

Operation and Quality Control

To ensure accuracy in flame photometric measurements, the optimum fuel-air ratio and sample aspiration rate should be held constant. Any change in the characteristic conal region of the flame should alert the analyst to a change in the fuel-air ratio. Maintenance of an optimal sample aspiration rate should be verified by timing accurately (ie, with a stopwatch) the volume of fluid collected per minute. The aspiration rate can be affected by the clogging of the atomizer due to the buildup of protein or particulate matter. It is important that the wire used to clean the atomizer does not change the size of the aspirator bore, because the flow rate would be affected.

Standards should be prepared with high purity salts in double-distilled or deionized water. Borosilicate glassware is satisfactory for short term storage of standards, but polyethylene containers are preferable for long term storage. Finally, an awareness of several possible sources of contamination, such as traces of residual soap on the analyst's fingers, cigarette smoke, and nail polish, should minimize such errors and ensure accuracy of measurements.

Limitations

Flame emission is a difficult process even for alkali metals with low ionization potentials. For example, in the measurement of sodium, only 1% to 5% of atoms reaching the flame are excited to emission. For the routine clinical measurement of levels of sodium, potassium, and lithium, however, the sensitivity of flame emission photometric measurements is adequate. Flame emission photometry is limited in its ability to analyze metals that are more difficult to excite. Fortunately, such elements can be readily analyzed by atomic absorption spectroscopy.

References

1. Willard HH, Merritt LL Jr, Dean TA, Settel FA: Instrumental Methods of Analysis. ed 6. New York, Van Nostrand, 1981, pp 127–153.
2. Henry RJ: Clinical Chemistry. Principles and Technics. New York, Harper & Row, 1967, pp 48–63.

5
Atomic Absorption Spectroscopy

Atoms can absorb energy of the specific wavelength corresponding to the wavelength they would emit when excited by a high-energy source. This property is utilized in atomic absorption spectroscopy. In atomic absorption, the energy required for atomic emission is supplied by a source other than a flame. The flame in atomic absorption serves as a cuvette to hold the majority of atoms in the sample in the ground state ready to absorb emission characteristic of the atoms provided by the emission source. The decrease in emission intensity of the source is proportional to the concentration of atoms of interest in the sample.

In Chapter 4, it was noted that flame emission is a difficult process. In contrast to flame photometry, there is a relatively greater number of atoms in the ground state than in the excited state. Therefore, with efficient atomization techniques, it is possible to maintain a relatively large population of atoms in the ground state that are able to absorb their characteristic emission wavelength from the source and contribute to the greater sensitivity of the measurement. Atomic absorption becomes the method of choice when the ratio of atoms in the ground state to those in the excited state is 1 billion or more, as is the case with zinc, which is virtually impossible to measure by flame photometry.

Design and Components

The basic components of a flame atomic absorption spectrophotometer are: an emission source, fuel mixtures for generating a flame, an atomization device, a nebulizer-burner, a monochromator, and a detector.[1] Atomic absorption spectrophotometers utilizing modified flame systems increase measurement sensitivity. The greatest sensitivity is provided by flameless or electrothermal atomic absorption instruments.

Figure 5–1

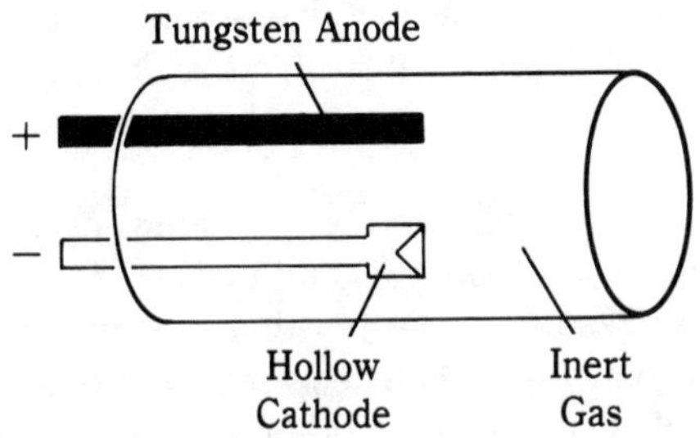

Hollow cathode lamp.

Source

To measure a particular element in a sample, the source must provide an emission wavelength characteristic of that element; that is, the source must contain the particular element and be capable of exciting it. This is accomplished in an evacuated glass envelope called a hollow cathode lamp (Figure 5–1). The element of interest or an alloy of the element is concentrated in a hollow, which serves as the cathode. Facing the cathode is an anode usually made of tungsten wire.

The lamp contains an inert gas, either argon or neon, that is under reduced pressure (approximately 1 to 5 torr). A potential is applied across the electrodes to produce a current of 1 to 50 mA; the inert gas is ionized at the anode, and positively charged ions speed toward the cathode. The impact of the ionized inert gas at the cathode dislodges the metallic element, which causes "sputtering." The sputtered metal atoms are excited by collisions with energetic inert gas ions and electrons and emit characteristic spectral lines that are absorbed by the corresponding atoms in the sample. Argon is preferred over neon in this arrangement because its greater atomic weight provides a greater impact on collision with the hollow metallic cathode and thus facilitates sputtering. In addition, argon has a lower ionization potential than neon and favors the emission of excited neutral atoms with the resultant intensity of emission of the desired spectral line. To isolate spectral lines in the ultraviolet region (wavelengths of less than 254 nm), the glass window of the hollow cathode lamp is replaced with quartz.

The hollow cathode lamp can also be improved by incorporating two additional electrodes coated with an easily excited species. The effect is to ionize additional inert gas atoms and increase the intensity of impact on the cathode resulting in spectral emission lines of increased intensity. These high-intensity cathode lamps, however, require an additional power supply. The hollow cathode lamp is adequate for most of the elements that can be determined by atomic absorption.

For volatile elements requiring sources of higher intensity, electrodeless discharge lamps are preferred. Made of silica tubing, these lamps, or quartz bulbs, are sealed with an inert gas such as argon at low pressure, together with the element of interest and small amounts of iodine or iodide salt, which is more volatile.

The quartz bulb is inserted in the cavity of a ceramic cylinder, over which is wrapped a radiofrequency (microwave) generator. The radiofrequency field ionizes the inert gas argon, which in turn excites the metallic atoms inside the lamp to emission. An increase in intensity 100 times greater than that in conventional hollow cathode lamps can be achieved with electrodeless discharge lamps that, however, require a special power supply.

For the determination of alkali metals (rubidium, cesium, sodium, and potassium) and other elements (zinc, cadmium, and mercury), a vapor discharge lamp is preferred. These lamps utilize metallic vapor that is excited to emission by a passing current. The lamp gas, usually neon, collides with electrons released during current flow, thereby exciting alkali metals and other elements that have low ionization potentials. The pronounced intensity of such lamps is due to the concentration of metal atoms and current in the lamp. Vapor discharge lamps require special mounts and power supply.

Fuel Mixtures

In flame atomic absorption spectroscopy, the function of the flame is to serve as a cuvette to hold the sample and, by the heat it generates, to break down molecules and maintain most of the resulting atoms in the ground state. In general, the air-acetylene flame, which reaches a temperature of 2400°C, is adequate for the atomization of most elements. However, elements that form refractory oxides need a higher temperature for conversion to atoms. This is achieved by using a nitrous oxide-acetylene flame, which reaches a temperature of 2800°C. Aluminum, germanium, and tin are among the elements that require determination by nitrous oxide-acetylene flame. The atomization of such elements can be increased by raising the temperature, but it is likely that the population of neutral atoms available for absorption will be reduced because ionization is favored. Just as in flame photometry, in which neutral atoms are excited and emit radiation of a characteristic wavelength, in atomic absorption it is the neutral atom that absorbs the emission from the hollow cathode lamp.

Nebulizer-Burner

As in flame photometry, the atomic absorption sample should be reduced to an aerosol or atomized mist as it enters the flame, a process accomplished in atomic absorption by using a premix or Lundegardh burner (Figure 5–2). In this arrangement, the sample is aspirated at a controlled flow rate through a nebulizer. The reduced air pressure created by the expanding oxidant entering through the nebulizer side arm draws the sample in the form of fine droplets that are then mixed with fuel and oxidant and carried to the burner head. Baffles or flow spoilers in the mixing chamber intercept large droplets, which are drained to waste; only about 10% of the sample reaches the flame. The drain utilizes a liquid trap to prevent combustion gases from reaching it.

The fuel/oxidant ratio in this system is controlled by an auxilliary oxidant inlet. The sample aspiration rate, which is controlled by the oxidant flow through the nebulizer side arm, is independent of adjustments made to control fuel oxidant ratio

Figure 5–2

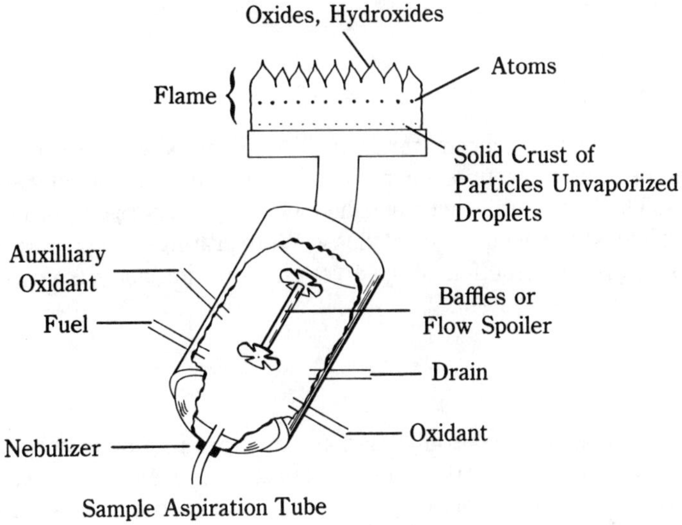

Premix nebulizer-burner system.

of the flame. Since the flame in atomic absorption also serves as a cuvette, the greater its path length, the greater the number of atoms it can accommodate, thus increasing the sensitivity of measurement. This is generally accomplished with single-slot, 10-cm burner heads that are used with air-acetylene flames. A triple-slot burner head is used when working with samples that contain a greater concentration of dissolved solids. Burner heads for nitrous oxide-acetylene flames use a 5-cm slot. The flame in the premix nebulizer-burner system has the shape of a fish tail, and is sometimes referred to as a "fishtail flame." Atomization is efficient because large droplets are drained to waste, resulting in a less "noisy" flame.

With this type of nebulizer-burner system, the following considerations exist:

1. The possibility of explosions in the mixing chamber. This possibility is minimized with the burning velocity of air-acetylene flames. Control units for igniting and extinguishing flames safely are used with nitrous oxide-acetylene flames, thus preventing the possibility of explosions.
2. Removal of combustion products generated in the flame. This is achieved by the exhaust hood positioned over the burner head.
3. Cleaning of the burner and mixing chamber. This necessary quality control function is complicated.

The flame itself is made up of several layers: unvaporized droplets at the bottom of the flame are overlaid with a crust of solid particles. Immediately above is a cloud of dissociated atoms, which are diluted by flame gases and surrounded in the outermost layer by oxides and hydroxides. The distribution of atoms in the flame varies for each element, and for this reason the height of the burner should be adjusted so that the source (hollow cathode lamp) is aligned with the optimum location of atoms in the flame.

Monochromator

The purpose of the monochromator is to isolate the wavelength of spectral line issuing from the emission source and resolve it from other overlapping and undesired wavelengths. This is generally accomplished with high-dispersion gratings that allow resolution of closely spaced spectral lines. With gratings wider slits can also be used so that a greater fraction of spectral emission intensity from the source is resolved by the optical system.

Detector

Photomultiplier tubes are generally used as detectors in atomic absorption instruments. Amplified signals are the pulsed signals of the emission source that are isolated by the monochromator, which ignores nonpulsed events such as emission in the flame and molecular spectra. Since the intensity of the desired spectral line reaching the detector is inversely proportional to the concentration of the measured element, low concentrations translate to smaller decreases in percent transmittance. Scale expansion could be used in such instances, provided that detector-amplifier noise is maintained at a negligible level by preventing nonpulsed emission of the flame reaching the detector-amplifier. The availability of microprocessors has made the task of computing transient signals from flameless instruments and the calibration and computation of concentration from linear and even nonlinear absorption-concentration plots relatively simple. It has also permitted automation of atomic absorption measurements.

Modified Flame and Flameless Systems

Among the disadvantages of the premix nebulizer-burner arrangement are the large sample size needed to compensate for dilution before aspiration and the fact that only 10% of the aspirated sample passes the flow spoiler and reaches the flame. This limits the sensitivity of the measurement of trace elements in biological fluids.

One approach to eliminate sample waste is the Delves technique, in which the sample is introduced directly into the flame. In this technique, a sample, such as whole blood, is dispensed into a nickel cup or a tantalum "boat." After the drying and digestion of organic matter with acid, the sample is introduced with an attachment arm either directly into the flame or underneath a slot of an elongated absorption tube in the flame. The latter arrangement used with cups spreads the atomized vapor through the absorption tube, which is aligned with the emission source. The flame temperature dissociates atoms, which are then available to absorb the energy from the emission source. The Delves technique has been used successfully for the measurement of lead in blood.

Figure 5–3

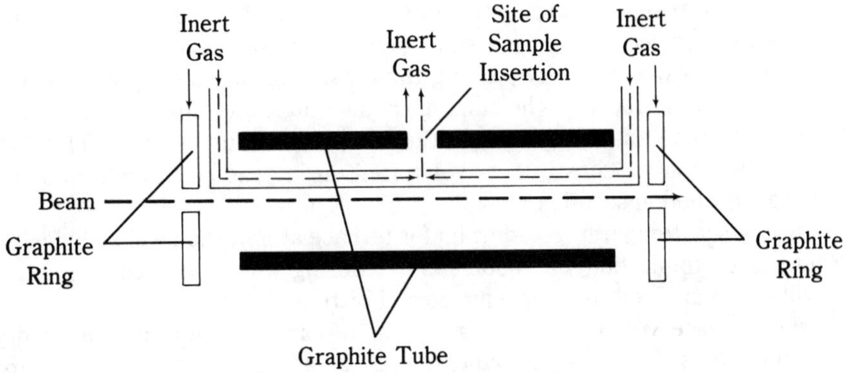

Graphite furnace—a flameless system.

Modified flame methods, such as the Delves technique, require sample preparation prior to flame analysis. Although these modified techniques have eliminated sample waste characteristic of the premix nebulizer-burner systems, they still are subject to dilution by flame gases, thus slightly affecting the sensitivity of measurement. The flameless systems are free from the dilution effects of flame and eliminate separate sample preparation steps required prior to analysis by modified flame systems. Although flame systems are rapid and have better precision, flameless systems offer the convenience of working with microliter quantities of sample and have the sensitivity of the order of nanograms to picograms.

An example of a widely used flameless system is the graphite, or Massmann, furnace, which consists of a graphite tube that is aligned in the path of the emission source (Figure 5–3). In the center of the tube is a receptacle for the insertion of the sample. At each end of the graphite tube is a graphite ring through which electrical contact is made. The sample is heated in three stages after insertion in the graphite tube. In the first stage the sample is dried at a temperature of approximately 100°C. The temperature is increased at a slow and controlled rate in the second stage to char, dry, and ash the sample. The charring step is designed to volatilize matrix constituents without volatilizing the element of interest. In practice, however, the temperature required to volatilize all of the matrix constituents may also be high enough to volatilize some of the test element. For this reason, charring temperature should be chosen judiciously; the optimal temperature may not be high enough for the removal of all of the matrix constituents yet will be low enough to prevent volatilization and loss of the test element. After the drying and charring steps, the temperature is rapidly increased in one step to effect atomization of the desired element. The atomic vapor produced absorbs the emission line from the source and generates a transient peak signal that can be recorded on a strip chart recorder or translated into a microprocessor readout of either peak height or peak area.

The graphite furnace is cooled by circulating water, which prevents radiation of heat to the environment and also permits rapid cooling of the tube after each atomization step to enable introduction of the next sample. The high temperatures attained in the furnace present a risk of degradation of the graphite tube by air oxidation, which is overcome by purging the outside of the tube with an inert gas. Inert gas is also introduced at the start of the heating cycle through the open ends of the graphite tube and exits through the sample introduction port. This flow of inert gas expels matrix constituents volatilized during the charring step and thus reduces background absorption.

Thus a typical temperature program for furnace atomization will include drying, charring, rapid atomization, and cooling steps during which, at each step, temperature will be maintained for a specific period of time.

During furnace volatilization, loss of analyte can occur during sample drying and charring steps. For volatile elements such as arsenic, selenium, and mercury this is more problematic. Zinc and cadmium can be lost during dry ashing at too high a temperature. In presence of chloride, loss of analyte can occur through volatile salt formation. The actual furnace temperature conditions chosen represents a compromise based on the characteristics of the analyte.

Problems inherent in the choice of sampling temperature are illustrated by the measurement of copper. Charring temperature of 950°C causes loss of copper. However, at temperatures less than 800° and at charring times less than 30 seconds, background absorption is increased. So a compromise charring temperature of 900°C and atomization temperature of 2600°C are used to produce the greatest sensitivity.

To minimize analytical problems in furnace atomic absorption spectroscopy, optimum charring temperature and drying time should be established. A surfactant such as Triton X-100 should be included with the sample to avoid foaming and spattering of serum during the drying step, especially in the analysis of chromium and manganese. Matrix modifiers should be added to serum where appropriate to stabilize element and to allow use of higher charring temperatures.[2] Thus magnesium nitrate can be used to stabilize aluminum, and nickel nitrate can be used to stabilize selenium. Sodium fluoride can be used to increase the volatility of molybdenum and to increase the sensitivity of the molybdenum signal.

The reason for the addition of matrix modifiers is that they decrease volatility of analyte element and prevent its loss during thermal pretreatment. Matrix modification allows application of higher pretreatment temperature for optimal matrix elimination and the use of a smaller temperature stop between pretreatment and atomization steps, which is important for reaching thermal equilibrium in the furnace. Matrix modification increases matrix volatility and removal of matrix constituents prior to atomization of analyte. Recently the use of a mixture of palladium and magnesium nitrates as matrix modifiers has permitted pretreatment temperatures of 900°–1400°C, thus permitting an optimized atomization temperature of around 2000°C to selectively volatilize the analyte of interest.

Flameless atomic absorption is facilitated by the use of a L'vov's platform, which is named after a Russian investigator who used a resistively heated graphite tube to atomize the sample. The advantages of using a L'vov platform are manifold. It permits atomization of the sample not from the wall of the tube, but from a

Figure 5–4

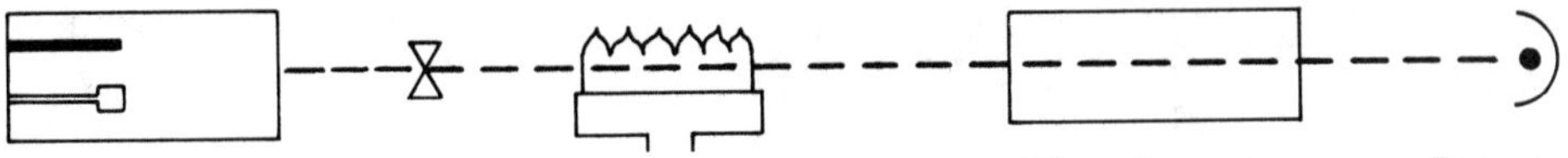

Single-beam atomic absorption spectrophotometer.

massive pyrolytic graphite inserted into the graphite tube. This delays atomization of the sample until the graphite and inert gas have reached thermal equilibrium. The platform has become an essential component of stabilized temperature platform furnace. The use of platform results in a longer tube lifetime since the sample does not come into contact with the tube wall.

The stabilized temperature platform furnace using the L'vov platform permits the temperature of inert gas in tube to be in equilibrium with the temperature of the tube wall. Under these conditions, formation of free atoms is optimum, and recombination of atoms to molecules with the resulting loss of atoms is effectively minimized. By completely interrupting inert gas during atomization, stabilized gas atmosphere within the graphite tube is achieved, while unnecessary cool down of the atom cloud is avoided, and loss of atoms from the tube is prevented. Additionally the use of peak area integration eliminates residual differences in atomization characteristics between sample and reference solutions.

Single - and Double - Beam Instruments

In a single-beam arrangement, emission from a source lamp is modulated with a rotating chopper or is pulsed electronically to provide an AC signal (Figure 5–4). The pulsed emission signal corresponding to the spectral line of the test element passes through a flame or furnace, and the monochromator isolates the desired spectral line. The intensity of pulsed emission remaining after absorption by test element is detected by a photomultiplier tube, amplified, and displayed on a readout as a concentration signal. Random emission from the flame is not amplified, thus permitting a selective measurement of absorption of source emission by atoms.

The drawback with the single-beam arrangement is that fluctuations in the intensity of the emission lamp during measurement are not compensated. In addition, a warm-up time is required to ensure that lamps have reached optimum and constant emission before measurements are made. These two drawbacks are overcome in a double-beam arrangement (Figure 5–5). In the double-beam arrangement, a portion of the emission signal from the source lamp bypasses the path of atoms in the flame or furnace and serves as a reference beam. Since the ratio of sample beams to reference beams is computed, any fluctuations in source intensity during measurement affect both the reference and sample beams and are canceled out. The double-beam arrangement permits use of lamps without prior warm-up and

Figure 5–5

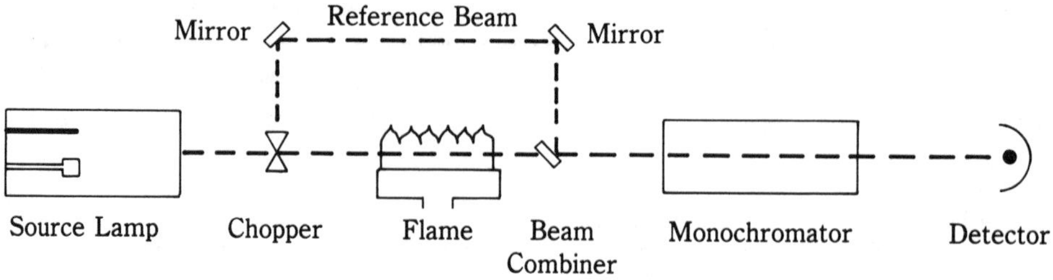

Double-beam atomic absorption spectrophotometer.

Figure 5–6

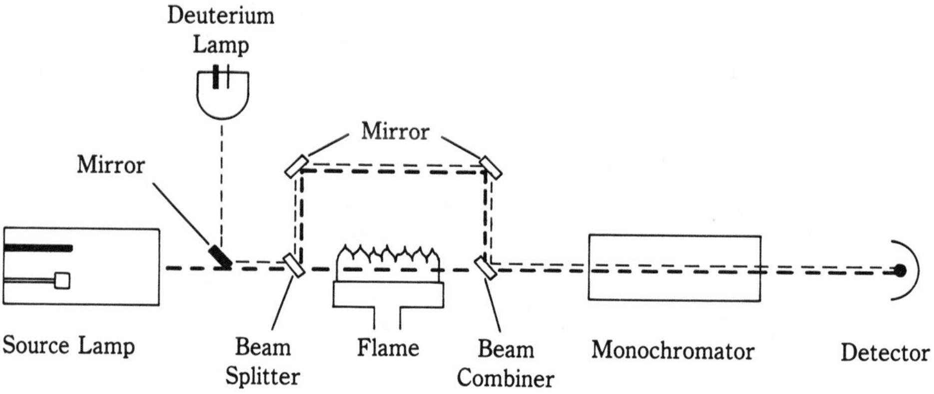

Atomic absorption spectrophotometer with background correction.

provides a stable baseline, making possible scale expansions of weak signals corresponding to low concentrations. The double-beam arrangement does not, however, correct for fluctuations in flame background. This can be achieved by using a continuum source, such as a deuterium arc lamp in the ultraviolet region or a tungsten-iodide lamp for the visible spectrum. In a double-beam arrangement with background correction, light from the emission and continuum source are split into reference and sample beams (Figure 5–6). The absorption of continuum source radiation by the test element is negligible. Background absorption, however, will be equal for both the emission and continuum source. When the ratio of the source emission to continuum emission is computed, background absorption is compensated.

In addition, the so-called Zeeman effect has recently found applications in background correction.[3] A magnetic field is utilized to split the emission wavelength of a metal into its polarized components. With this approach, the background signal is subtracted from the sum of both background and analyte signals. Background correction is particularly important in flameless systems in which background absorption caused by smoke and matrix constituents can lead to erroneously high values.

The Zeeman effect relies on the fact that the application of a magnetic field causes an atomic spectral line that is either emitted or absorbed by an atom to be split into three or more polarized components.[3] The π component is found at the original wavelength of the spectral line, while the wavelength of the other two components, the σ components, is shifted equally on either side of the wavelength of the original line. The π component is polarized in a plane that is parallel to the magnetic field, whereas the σ components are polarized in a plane that is perpendicular to the magnetic field. This pattern is called a "normal" Zeeman pattern and is caused by atoms such as zinc and cadmium. Most atoms in fact have a complicated or "anomalous" pattern in which there exists a splitting of π and σ components. The Zeeman effect can be exploited to correct background absorption because molecules and solid particles that contribute to background absorption do not evidence Zeeman splitting as atoms do. Thus when the magnetic field is not applied, both the background and the absorption resulting from the test atom are measured. When the magnetic field is applied to the atomic vapor in the graphite furnace, only the background is measured, because the σ components are shifted away from the atomic emission line, while the π component at the original wavelength of the atomic emission line is ignored by the polarizer. Thus only the background is measured with the magnetic field on. The difference between the signal obtained when the magnetic field is off and the signal obtained when the magnetic field is applied corresponds purely to the signal given by the test element.

In place of a continuum source, one can employ a second source emission lamp in the double-beam arrangement.[1] Such a two-channel, double-beam instrument will have two monochromators, two detectors, and a readout. Computing the ratio of pulsed signals from the two lamps will effectively compensate for background absorption. Alternatively, the second lamp can be used for the simultaneous determination of another element in addition to the element measured with the primary source lamp. The same element can be measured using two emission lines of different sensitivity, one from each lamp, thus extending the analytical range of measurement.

Interferences

Interferences in atomic absorption measurements can arise because of the sample matrix and chemical and ionization reactions.

Matrix

Matrix interferences arise when the standard and sample solutions have differing viscosities or surface tensions and are not atomized at the same rate. The effect of viscosity is overcome by making up the standards in the same matrix as the sample or by mimicking the viscosity of the sample matrix. Samples with high salt concentrations require energy to decompose it and dissipate the energy required to atomize the sample, leading to apparently lower values. Such samples require dilution. Matrix interferences can be overcome by the method of additions, in which the measurement of the sample is followed by measurement of two or three aliquots of samples containing increasing concentrations of standard. If the plot of absorbances of the sample and matrix standards are linear, the plot can be extrapolated to the x-axis to read the true concentration of the sample.

Chemical

Chemical interferences result if the element of interest exists in the form of a thermally stable compound and is not readily dissociated into neutral atoms (eg, calcium and magnesium form thermally stable complexes with phosphate). To dissociate these complexes, an excess of an element such as lanthanum or strontium is added to form a firmer complex with phosphate and thus release calcium and magnesium for measurement. Chemical interferences can also be overcome by using a hotter flame, such as a nitrous oxide-acetylene flame; this is especially effective for dissociating refractory oxides of metals, but the possibility of ionization makes this method less attractive. Chemical interferences are minimal in the inert gas atmosphere of flameless systems.

Ionization

Ionization reduces the population of neutral atoms available for absorption and measurement. With hotter flames, ionization interference can become a serious problem affecting the sensitivity of the measurement. Just as in flame photometry, ionization of the measured element can be suppressed by adding in excess an easily ionized element, such as strontium, cesium, or potassium, to both standard and sample. The large number of free electrons resulting from the ionization of the additional element effectively suppresses the ionization of the measured element.

Sampling Techniques

Analysis of biological fluids, such as serum or plasma, is usually done either after appropriate dilution with buffer in flame systems or without pretreatment in flameless systems. Metal bound to protein can be released after precipitation of protein with trichloroacetic acid. The metal is then chelated in the protein-free filtrate with a chelating agent, such as ammonium pyrrollidone dithiocarbamate, after appropriate adjustment of pH, and extracted with an organic solvent, such as methyl isobutyl ketone, prior to measurement. This extraction procedure is limited by the multiple steps involved and the possible loss in recovery at each step.

Sensitivity and Detection Limit

Sensitivity in flame systems is expressed as the concentration of an element in micrograms per milliliter that will produce an absorbance of 0.0044 or 1% absorption. For flameless systems, sensitivity is stated in terms of weight in picograms of element that will produce 1% absorption. Verification of the stated sensitivity of an element for an instrument ensures that one has optimized the working parameters and the instrument is performing satisfactorily.

Detection limit is the concentration that can be distinguished from baseline noise and is defined for flame systems as the concentration that will produce a signal to noise ratio of 2 to 1. For flameless systems detection limit is stated in terms of the weight of the element that will produce the same ratio.

Operation and Quality Control

For each element, the alignment of both the lamp and the burner must be adjusted to produce an optimum absorption signal. The sample aspiration or nebulization rate should also be adjusted in flame systems to obtain optimum absorption.

The process of "sputtering" in the lamp during measurement eventually reduces the lifetime of the lamp. Thus, lamps designed for measuring volatile elements, such as selenium, cadmium, and arsenic have a shorter lifetime. Evolution of hydrogen by some lamps will produce background emission and affect the sensitivity of the measurement. This can be overcome by reversing the current polarity for a few minutes during which time the hydrogen gas will be removed by adsorption to a tantalum-getter material attached to the anode. Lamps eventually fail because of loss of carrier gas, which becomes absorbed to the inner surface of the lamp in a process referred to as "gas clean up." Lamp life is extended by using larger lamps, which can accommodate more inert gas.

Applications

Atomic absorption spectroscopy has permitted analysis of clinically important trace metals.[4] Analysis of cadmium, lead, and arsenic by sensitive flameless techniques has provided a new dimension to the determination of toxicologically important trace metals. The ability to measure zinc, selenium, manganese, and other nutritionally important trace elements has provided a new impetus for this research field.[5] The investigation of aluminum toxicity in patients with chronic renal failure and knowledge of value of serum copper in the investigation of specific diseases of pediatric patients have emerged with the availability of the powerful analytical tool of atomic absorption.

References

1. Willard HH, Merritt LL, Jr, Dean TA, Settle FA: Instrumental Methods of Analysis, ed 6. New York, Van Nostrand, 1981, pp 127–153.
2. Ericson SP, McHalsky ML, Rabinow BE, et al: Sampling and analysis techniques for monitoring serum for trace elements. Clin Chem 32: 1350, 1986.

3. Brown SD: Zeeman effect-based background correction in atomic absorption spectrometry. Anal Chem 49: 1269A, 1977.
4. Mertz W: Trace element nutrition in health and disease: Contributions and problems of analysis. Clin Chem 21:468–475, 1975.
5. Sunderman FW Jr: Electrothermal atomic absorption spectrometry of trace metals in biological fluids. Ann Clin Lab Sci 5:421–434, 1975.

6
pH Measurements by Potentiometry

A resurgence of electroanalytical methods in the clinical laboratory is evidenced by the incorporation of such methods to discrete and continuous-flow, fast analyzers. The measurement in an electrochemical cell of such electrical parameters as potential, current, and resistance forms the basis of electroanalytical methods. These parameters can be measured as a function of time and can be grouped into several categories such as potentiometry, coulometry, voltammetry, and conductometry. Potentiometry involves measurement of potential (E) at zero current flow. Coulometry utilizes measurement, in coulombs, of the electric current (Q) generated at an electrode by an electroactive species. This measurement is dependent on the flow of current (I) in amperes through the cell during a time interval (T) in seconds ($Q = IT$). Coulometry is the basis of Cotlove's chloridometer, which is used for the electrometric determination of chloride. Voltammetry involves measurement of the current generated when a voltage or potential is applied to a cell. Voltammetric techniques include amperometry and polarography, the former being employed in electrometric titrations, and the latter being the basis of oxygen-sensing electrodes used for measurement of PO_2 (partial pressure of oxygen) in blood, and enzymatic reactions utilizing oxygen (glucose oxidase and cholesterol oxidase for the determination of glucose and cholesterol, respectively). The measurement of resistance or the reciprocal of resistance (conductance) is the basis of conductometric measurements, an obvious application of which being the determination of the purity of distilled water.

Concept of Potential

When a metal is immersed in a solution of its own salt, such as a zinc rod in a solution of zinc sulfate, the metal will begin to dissolve, thereby releasing electrons

($Zn^0 \longrightarrow Zn^{++} + 2e$), and will be more negatively charged relative to the solution interface. The release of electrons by zinc metal amounts to oxidation occurring on its surface, and the zinc metal or electrode can be characterized as the anode. The zinc/zinc sulfate system represents a half cell. If this half cell is in contact through a porous barrier or salt bridge with another half cell, such as copper metal immersed in a solution of copper sulfate, and if a wire is connected to the zinc and copper electrodes, the electrons from the zinc electrode will flow to the copper electrode, and the reaction ($Cu^{++} + 2e \longrightarrow Cu^0$) will occur. When copper is deposited on the copper electrode, it will represent the cathode, since reduction or gain of electrons takes place. The porous barrier or salt bridge connecting the zinc and copper electrode systems prevents mixing of the two solutions and free movement of ions between the two compartments to maintain electrical neutrality. The liquid junction potential that develops when two different solutions are in intimate contact is also minimized.

In this electrochemical cell constituted of zinc and copper half cells, the reactions occur spontaneously when electrons flow through a wire. This type of electrochemical cell is called a "galvanic" or "voltaic" cell, in contrast to an electrolytic cell, in which external voltage in excess of the reversible galvanic cell voltage has to be applied for chemical reactions to occur.

The half cell reactions occurring in the zinc and copper electrochemical galvanic cells can be depicted as

$$Zn \mid Zn^{++} \parallel Cu^{++} \mid Cu$$

The single vertical lines indicate phase boundaries where potential differences exist, and the double vertical lines indicate contact between solutions in the half cells by means of a porous barrier or salt bridge (liquid junction). By convention, electrochemical reactions are written with the oxidation reactions on the left and reduction reactions on the right.

Electrode Potential

It is not possible to determine the potential of a half cell or an electrode since an electrochemical cell is composed of two half cells. The cell potential or electromotive force is the algebraic sum of the potentials of the two half cells. A standard hydrogen electrode has been arbitrarily assigned a potential of zero volts at 25°C, permitting comparison of potentials of other half cells. Thus, the electromotive force of a cell made up a half cell (cathode) in combination with a standard hydrogen electrode (anode) represents the potential (E) of the unknown half cell, assuming that the contribution of the liquid junction potential, which is minimized by a salt bridge, is zero, since

$$E_{cell} = E_{cathode} + E_{liq\ junc} - E_{anode}$$

Two limiting factors exist in determining the electrode potential of a half cell relative to a reference half cell: one is the solution pressure of the electrode metal, which is dictated by the ability of the metal to ionize; the other is the effective concentration or the activity of the dissolved ionic species of the metal, which at constant temperature varies with concentration.

The Nernst equation is used to calculate the true potential of a half cell when the activities of the ions in the two half cells are not the same:

$$E = E^0 - \frac{2.303RT}{nF} \log \frac{A_{red}}{A_{ox}}$$

where E is the corrected electrode potential, E^0 = standard electrode potential (potential when activities of ions at 25°C are at unity), R = molar gas constant (8.314 J/mol), T = temperature (°K), n = number of moles of electrons transferred, F = Faraday's constant (96,500 C per equivalent weight), A_{red} = activity of the reduced species, and A_{ox} = activity of the oxidized species.

The Nernst equation at 25°C (298°K) simplifies to

$$E = E^0 - \frac{0.05916}{n} \log \frac{A_{red}}{A_{ox}}$$

From this equation, it is apparent that a change by one unit in the logarithmic term will shift the electrode potential by 0.05916/n V or 59.16/n mV.

Relationship of Potential to pH

By definition, pH is the negative logarithm of hydrogen ion concentration or activity: pH = $-\log [H^+]$. The indicating electrode is used to measure $[H^+]$ in a solution and constitutes a half cell. In addition, a reference electrode is needed to determine the potential of the indicating electrode. If one assumes that a hydrogen gas electrode is used as the indicating electrode and a silver/silver chloride electrode as the reference electrode (the electrode combination originally used by the National Bureau of Standards to establish standards for calibrating pH meters), the following half cell reactions are obtained:

$$H_2 \rightarrow 2H + 2e \ (E^0 = O\ V)$$
$$2AgCl + 2e \rightarrow 2Ag^0 + 2Cl^-$$

The total reaction can be expressed as

$$H_2 + 2AgCl \rightarrow 2H^+ + 2Ag^0 + 2Cl^-$$

yielding E^0 = 0.222 V. Applying the Nernst equation to the above reaction,

$$E = E^0 - \frac{0.05916}{n} \log \frac{A_{red}}{A_{ox}} =$$

$$0.222 - \frac{0.05916}{2} \log \frac{[H^+]^2[Ag^0]^2\,[Cl^-]^2}{[H_2]\,[AgCl]^2}$$

Assuming that gases and solids have an activity of 1 and the activity of $[Cl^-]$ to be unity, the equation reduces to

$$E = 0.222\ V - 0.5916 \log [H^+]$$

or

$$E = 0.22\ V + 0.05916\ pH$$
since $-\log [H^+]$ = pH.

Figure 6–1

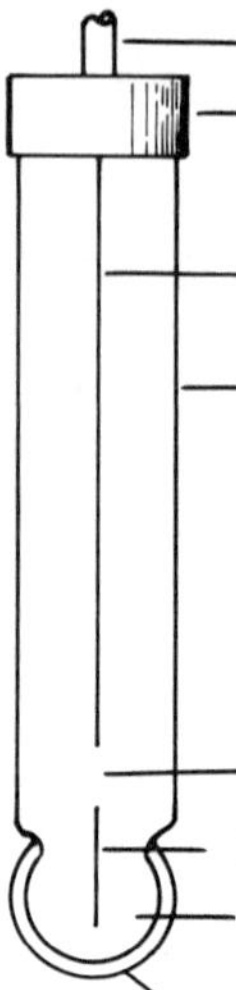

Glass electrode.

Thus, one can see the relationship between the potential developed across an indicating electrode and the pH of the solution in which this electrode is immersed; also, for every unit change in pH, the potential will vary by 59.16 mV.

Glass and Reference Electrodes

A glass electrode consisting of a glass bulb that is sensitive to H^+ activity is illustrated in Figure 6–1. The glass bulb can be made particularly selective to H^+ if the glass bulb contains lithium silicate with added barium and lanthanum ions.[2] Sealed onto the tapering glass bulb is a stem of high-resistance glass which itself is not sensitive to H^+. Inside the glass stem is an internal reference electrode, such as silver/silver chloride, dipping in dilute hydrochloric acid (0.1 N) and a buffer solution containing potassium chloride. The internal reference electrode is thus surrounded by a constant hydrogen ion concentration.

When the glass electrode is immersed in a solution, there is an exchange of lithium cations from the surface of lithium silicate glass for H^+ that are taken up by the silicate lattice. As a result, the outer surface of the glass bulb is transformed into a hydrated gel layer, that may be several hundred nanometers thick depending on the extent of the hygroscopicity of the glass layer. The ion exchange equilibrium can be represented by the equation:

$$\underset{\text{(solution)}}{H^+} + \underset{\text{(glass)}}{Li^+} \rightleftharpoons \underset{\text{(glass)}}{H^+} + \underset{\text{(solution)}}{Li^+}$$

Figure 6–2

Solution inside glass	Inner hydrated gel layer	"Dry" glass	Outer hydrated gel layer	Solution (pH to be measured)

Events occurring across the glass electrode.

It is apparent that as lithium from the glass dissolves into the solution in which the glass bulb is immersed, the exposed dry glass becomes hydrated to maintain a steady state on the outer glass surface. Consequently, the rate at which the glass surface is depleted will determine the lifetime of the electrode. In practice, the dry glass bulb of the glass electrode is of sufficient thickness (10^{-2} cm) to last for several years of continual depletion in solution. The exchange of H^+ ions from the solution whose pH is to be measured into the gel layer results in a potential difference across the boundary of the hydrated gel layer and the solution, the magnitude of which is proportional to H^+ activity. Since the inner surface of the glass bulb is also resting in an acid solution, an inner hydrated gel layer is created. Therefore, one must consider the potential of this layer and of the internal reference electrode, as well as the "asymmetry potential" that develops as a result of the strain in the glass created by the H^+ on the inside and outside of the pH-sensitive glass bulb. Again, the glass electrode represents a half cell whose potential cannot be measured without another half cell; this is accomplished with another reference electrode. The potential of the cell is measured as the sum of the potentials developed at the glass electrode (outer hydrated gel layer + inner hydrated gel layer + internal reference electrode) and at the reference electrode. Events occurring across the glass electrode are summarized in Figure 6–2.

Since the potential developed at the interface of the solution whose pH is to be measured and the outer hydrated gel layer is the only variable, and all other potentials are constant, the potential measured by the pH meter is directly related to the H^+ activity and thus the pH of the solution. In practice, the pH of an unknown solution is measured only after standardizing the pH meter with at least two standard buffers representing the range within which the pH of the measured solution is expected to fall.

Either the silver/silver chloride electrode or the calomel (mercury/mercurous chloride) electrode can be used as reference electrodes to measure potential developed at the glass electrode. Recently, the rapid response of thallium amalgam/thallous chloride reference electrodes has been found to present an advantage over silver and mercury electrodes when one is taking measurements over a wide range of temperatures. In addition, the reference electrode can be combined with the glass electrode to yield a compact combination electrode, thus dispensing with the need to use a separate reference electrode. In a combination electrode, the reference electrode is incorporated within the glass electrode assembly, and there is also an opening for the introduction of a saturated salt solution (potassium chloride)

Figure 6–3

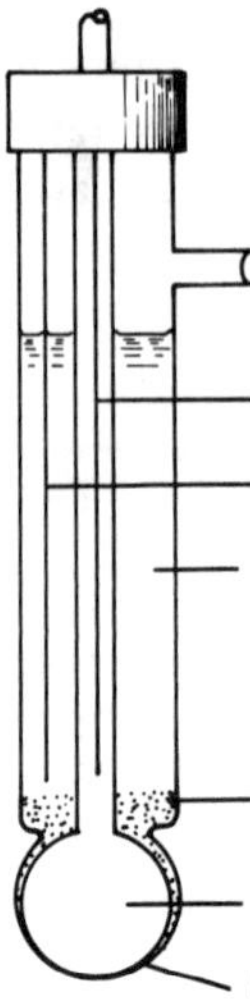

pH combination electrode.

that forms a salt bridge (Figure 6–3). In some electrodes, an asbestos fiber sealed onto a glass, ceramic, or porous glass plug can serve the function of a salt bridge.

Measurement of pH

The potential developed at the boundary of the unknown solution and the outer hydrated gel layer of the glass electrode is measured at zero or negligible current flow. Meters that read pH with an accuracy of 0.001 pH require high impedance voltmeters that draw a current of 10^{-12} A or less, so that pH measurements are performed at virtually zero current flow. Such pH meters can also compensate for temperature changes that would otherwise affect the measurement. Readings to 0.001 pH are possible because of scale expansion capabilities that allow a pH of 0.5 or 1.0 to be spread over the full meter scale.

Once the high impedance voltmeter type pH meter is standardized with two buffers that reflect the pH range within which the pH of the unknown solution is expected to lie, pH response of the unknown can be expected to follow the Nernst relationship and is determined accurately by interpolation. Direct-reading, high-impedance, voltmeter-type pH meters have largely supplanted the traditional null-balance potentiometer-type pH meters.

Practical Considerations

The glass electrode must be sufficiently hydrated before pH measurements are taken. Dry electrodes should be soaked in buffer for at least 24 hours to achieve

Figure 6–4

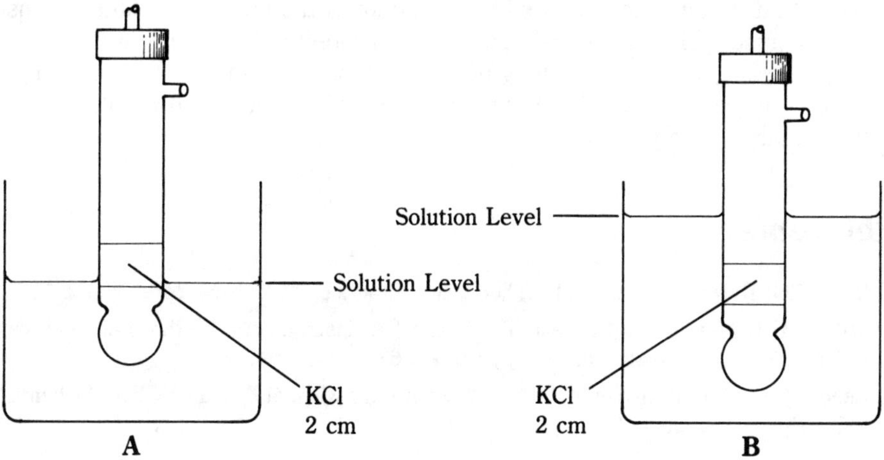

Immersion of electrodes. A: Correct depth of immersion–the salt bridge electrolyte is free flowing. B: Immersion too deep–the free flow of electrolyte is interrupted.

hydration. It is recommended that electrodes soak in buffer when not in use. Although the aperture for introducing the salt bridge electrolyte (KCl) is capped when not in use, the cap should be removed when measurements are performed to ensure free flow of electrolyte. This flow can be as small as 0.1 to 0.01 mL per day depending on the type of salt bridge junctions used. Interruption of flow of the salt bridge electrolyte will lead to incorrect results because of the lack of ion conduction between the electrodes of the cell.

The level to which the combination electrode is immersed in solution is critical and should be such as to result in a pressure head of about 2 cm to ensure free flow of salt bridge electrolyte.[3] Electrodes immersed too deeply in solution will result in contamination of the solution inside the glass electrode (internal solution) and yield incorrect results. Figure 6–4 illustrates the proper height to which electrodes should be immersed.

The accuracy of pH measurements must be insured by calibration with National Bureau of Standards (NBS) primary standard buffers, which are preferably within 1 pH unit of the pH of the unknown solution. This can be readily done by using one of the six NBS-certified buffers that are available in the pH range 3.5 to 10.5. The glass electrode is not suitable for measurements below pH 2 and above pH 12. In acidic media below pH 2, the H^+ activity of the hydrated gel layer is affected, leading to "acidic" errors. In contrast, in highly alkaline media, cations other than H^+ are exchanged between the hydrated gel layer and the solution, resulting in "alkaline" errors. If measurements are taken in either highly acidic or highly basic media the pH meter must be standardized with appropriate secondary standards such as potassium tetroxalate (pH 1.68 at 25°C) and calcium hydroxide (pH 12.45 at 25°C).

Because temperature affects pH values, it is important that buffers and the unknown solution are at the same temperature. The electrodes should be thoroughly rinsed with distilled water followed by an aliquot of the unknown solution to insure that the correct pH is measured. To prevent protein build-up on the pH-sensitive glass membrane, thorough washing of the electrode is mandatory after taking the pH of proteinaceous solutions. A solution whose pH is measured should be stirred well to obtain a true pH reading.

References

1. Bates RG: Determination of pH: Theory and Practice. ed 2. New York, Wiley, 1973.
2. Willard HH, Merritt LL Jr, Dean TA, Settle FA: Instrumental Methods of Analysis, ed 6. New York, Van Nostrand, 1981, pp 652–661.
3. Bender GT: Clinical Instrumentation: A Laboratory Manual Based on Clinical Chemistry. Philadelphia, Saunders, 1972, pp 195–207.

7
Ion-Selective Electrodes

In principle, an electrode can be made selective to measure the activity or effective concentration of an ion. In practice, however, such electrodes are rarely totally specific for a particular ion. The selectivity of an ion-selective electrode is expressed in terms of the selectivity constant (Kij), which is a measure of the selectivity of the measured ion (i) relative to the interfering ion (j). Thus the selectivity constant (Kij) of an ion-selective electrode for potassium (the valinomycin electrode) for the interfering ion sodium is 10^{-4}. This means that the potassium ion-selective electrode is 10,000 times more sensitive to potassium than sodium. A knowledge of the selectivity constant of an ion-selective electrode for interfering ions permits an assessment of selectivity of the ion-selective electrode.

Glass and Membrane Electrodes

The composition of ion-selective electrodes can range from a glass membrane, the composition of which renders it selective to a particular ion, to a solid-state or liquid-membrane electrode, or an electrode coated with enzyme.[1]

Glass electrodes can be made selective to a cation such as sodium by including sodium as a component of the glass membrane. Thus a typical sodium ion glass electrode membrane is composed of 71 parts of SiO_2, 11 parts of Na_2O, and 18 parts of Al_2O_3. Because hydrogen ions interfere, their activity should be kept relatively low with respect to the sodium ion by making measurements in well-buffered solutions to ensure that pH does not drift during measurement. The membrane electrodes can be classified as either solid-state or liquid-membrane. The solid-state membrane electrode can consist of a single crystal or pressed crystals of poorly soluble salts. In either case, the solid-state membrane electrode is referred to as homogeneous, (examples of which being a solid-state membrane electrode made of pressed polycrystalline pellets of silver chloride, or a mixture of silver chloride or silver sulfide). Such an electrode can be used to measure chloride ion. Unfortunately, sulfide, iodide, bromide, thiocyanate, and cyanate ions interfere with measurement. As such, in bromide intoxication, chloride values measured with the chloride ion-selective electrode will be overestimated.

A heterogeneous solid-state membrane electrode can be constructed by incorporating a poorly soluble precipitate of a salt in an inert matrix such as silicone rubber, polyethylene, or acrylamide. Thus a precipitate of silver chloride embedded in a matrix can be made selective to chloride ions, although it must be recognized that it will also measure interfering ions such as bromide.

A liquid-membrane electrode can be constructed with an ion-selective material dissolved in an organic solvent.[2] Thus a calcium-selective electrode can be made by dissolving a liquid ion-exchanger such as calcium salt of didecylphosphoric acid in an organic solvent such as di-n-octylphenyl phosphonate. This calcium ion electrode is also sensitive to hydrogen and zinc ions.

The fact that antibiotics form specific complexes with cations has been made use of in the design of the valinomycin electrode for the measurement of potassium.[3] The liquid membrane electrode consists of valinomycin in a matrix of diphenylether or dibutylsebacate. In theory, this electrode is more sensitive to rubidium than potassium. However, because serum is relatively free of rubidium, this limitation is of no practical consequence.

The recent availability of electrodes containing ion-specific materials cast in synthetic matrixes such as polyvinylchloride (PVC) has provided electrode membranes that are mechanically resistant and, as such, has extended the shelf life of electrodes. The lipophilicity of the membrane component has a bearing on its life time. The lipophilicity of PVC plasticizers such as Bis (1-butyl pentyl) decane-1, 10-diyl-diglutarate (ETH 469) used in PVC membrane-based ion-selective electrodes are responsible for their stability.

Ion-selective electrodes can be constructed by coating a glass electrode with an enzyme. Such an enzyme electrode can be used to measure a substrate such as urea, which can be measured by using a glass electrode coated with the enzyme urease. The potential developed at the electrode membrane surface by the ammonium ions produced by the action of urease on urea can be related to the urea concentration.

Relative Merits

Direct potentiometry on undiluted plasma or serum samples using ion-selective electrodes measures the activity or effective concentration of an ion in the sample. Flame photometry, however, measures the concentration of an ion present in the plasma compartment of the sample. Because normal plasma contains approximately 93% water, flame photometric measurements, compared to direct potentiometric measurements, should theoretically be 7% lower in normal plasma.[4] In practice, however, values obtained by flame photometry are approximately 2% to 4% lower than direct ion-selective electrode measurements. The difference has many causes, including the binding of an ion such as sodium to bicarbonate, and the binding of water to proteins that may increase the effective water content of the plasma sample to 96%. Some ion-selective electrode instruments have built-in corrections to account for normal protein levels in plasma thus giving agreement with flame photometry values. The problem arises, however, when the protein and lipid content of plasma exceeds normal levels. In such instances, flame photometric results of the concentration of an ion such as sodium will be, compared to direct potentiometric results, underestimated. A similar discrepancy will also be noted in indirect potentiometry measurements using diluted sample, because dilution will not take into account volume occupied by abnormal quantities of lipids and proteins. Thus the effect of lipemia or the presence of elevated protein levels, such as in multiple myeloma, can produce pseudohyponatremia by flame photometric and indirect potentiometric measurements when compared to direct potentiometric measurements.

Direct potentiometric ion-selective electrode measurements of the same sample on different direct potentiometric instruments may vary because of differences in calibrant composition and variations in residual liquid junction potential. The latter are induced by differences in reference electrode design and salt bridge composition. Thus when the concentration of potassium chloride used as a salt bridge is reduced, the measured values of ions such as sodium and potassium are also reduced.

Residual junction potential, which arises from differences between the junction potential with the sample and the calibrant, can be minimized by adjusting the composition of the calibrant to that of the sample. Liquid junction potential is increased by erythrocytes in the whole blood sample, compared to plasma, because the potassium chloride solution used as salt bridge crenates the red cells and precipitates proteins. The precipitated proteins act as an ion-exchanger slowing down chloride ions more than potassium ions thus creating a positive diffusion potential. A flowing junction minimizes the effect of erythrocytes. The effect of erythrocytes on the liquid junction potential can be rendered negligible by use of sodium formate as bridge solution. In such a case, whole blood values will correspond to values obtained with plasma.

Practical Considerations

Practical problems in ion-selective electrode measurements can be illustrated by difficulties in the measurement of ionized calcium. The amount of ionized calcium is dependent on the pH and protein concentration of the blood sample. Some of the calcium ion-selective electrode analyzers simultaneously measure both the calcium ion concentration and the blood pH. In this manner, such instruments are able to compute both the actual calcium ion activity and calcium ion activity corrected to a normal pH of 7.4.

To ensure accurate measurements of ionized calcium, the calibrator solutions used should have the same activity coefficient as calcium ion concentration in normal plasma. The composition of the calibrator solution should be such that it produces the same liquid junction potential as does plasma at the salt bridge interface. Buffers such as HEPES (*N*-2-hydroxyethylpiperazine-N^1-2-ethane sulfonic acid) used in calibrators bind calcium strongly, leading to a decrease in ionized calcium. An alternative to HEPES is the use of a buffer such as MOPS (3-(*N*-Morpholino) propane sulfonic acid) that exhibits negligible calcium binding. Ionic calcium measurements are subject to interference by other cations such as magnesium. Interference can be compensated by adjusting the composition of the calibrator solution to include the interfering cations. There is confusion regarding the interference of albumin with the direct potentiometric measurements of ionized calcium. A recent study was unable to demonstrate the in vitro interference of albumin on ionized calcium measurements.[5] Finally, specimen collection requirements for ionized calcium requirements are stringent.[6] Ideally pH variations in whole blood can be minimized by preventing loss of carbon dioxide through anaerobic sampling. The specimen should be processed as rapidly as possible and maintained in ice to minimize the formation of lactic acid. If blood is collected in closed tubes, the tubes must be completely filled to eliminate any headspace dilution. The ideal anticoagulant is heparin, preferably calcium-titrated. Thus it is necessary to be aware not only of variables in ion-selective electrode measurements, but also of variations due to specimen collection, stability, and handling.

References

1. Freiser H: Ion-selective Electrodes in Analytical Chemistry, vol 1 (1978), vol 1 (1980). New York, Plenum Press.
2. Oesch U, Amman D, Simon W: Use of ISE's for clinical analysis: past and future. Methodology and clinical applications of ion-selective electrodes. 7:273–279, 1986. (IFCC workshop, Radiometer, Copenhagen.)
3. Oesch U, Amman D, Simon W: Ion-selective membrane electrodes for clinical use. Clin Chem 32:1448, 1986.
4. Apple FR, Koch DD, Graves S, Ladenson HJ: Relationship between direct potentiometric and flame photometric measurement of sodium in blood. Clin Chem 28:1931–1935, 1982.
5. Thode J, Boyd JC, Ladenson JH: Activity measurements of calcium, sodium, potassium, and chloride after equilibrium dialysis used to show lack of evidence for protein interference with calcium electrodes. Clin Chem 33:1811, 1987.
6. Boink ABTJ, Buckley BM, Christiansen TF, et al: Recommendations on blood sampling, transport and storage for the determination of the substance concentration of ionized calcium. Methodology and clinical applications of ion-selective electrodes. 8:81, 1987 (IFCC workshop, Radiometer, Copenhagen.)

8
Coulometry, Polarography, and Anodic Stripping Voltammetry

Coulometry

Coulometric methods measure, in coulombs, the quantity of electricity generated in a reaction. Since 96,487 coulombs (Faraday's constant [F]) is needed to liberate one equivalent weight of a substance at an electrode during a coulometric titration, the measurement of the number of coulombs (Q) can be related to the equivalent weights (W) of the electroactive species in a sample. Mathematically, this relationship can be expressed by Faraday's law: $W = Q/F$. The number of coulombs generated in a reaction is related to the product of current in amperes (I) that flows through the cell and time in seconds (T) during that flow ($Q = I \times T$).

The equivalent weights of the electroactive species in the sample can then be related to the electricity generated in the reaction and Faraday's constant ($W = IT/F$). Multiplying the number of equivalents by the equivalent weight (molecular weight [M]/valence [n]) gives the concentration (c) of the electroactive species:

$$c = (IT/F)(M/n)$$

Applications

In the clinical laboratory, coulometry is used for the measurement of chloride: most coulometers are based on the principle of constant-current coulometry. Because the current is constant throughout the reaction, the number of coulombs generated can be related to the time required for the titration.

The coulometer used for the electrometric determination of chloride consists of a pair of silver generation electrodes and another pair of silver electrodes to amperometrically detect the end point of the titration. A timer is activated as soon as the sample is titrated with silver ions generated at the silver anode by application

of a fixed potential of 1 to 2 V. The sample is diluted with a mixture consisting of nitric acid, acetic acid, and gelatin, and provides for electrolytic conduction.[1,2] Silver ions generated at the anode react with chloride ions to form insoluble silver chloride. Nitric acid provides protons to facilitate the cathode reaction. Acetic acid, which decreases the polarity of the solution, ensures that silver chloride remains insoluble. The presence of gelatin in the reaction mixture improves reproducibility of titrations. The inclusion of thymol blue in the reagent mixture not only prevents microbial growth but also serves as a visual indicator for the presence of gelatin and acid by the solution remaining pink. The reactions at the silver generator electrodes can be summarized as follows:
at the anode,

$$Ag^0 \rightarrow Ag^+ + e^-$$
$$Ag^+ + Cl^- \rightarrow AgCl$$

at the cathode,

$$2H^+ + 2e^- \rightarrow H_2$$

The amperometric detection of the end point of the titration is achieved by the other pair of silver electrodes, to which a constant potential of 0.25 V is applied. As soon as all the chloride ions in the sample have reacted with silver ions, the subsequent generation of silver ions results in an increase in the current across the amperometric electrode. The increase in current triggers a relay mechanism, which stops the timer and the generation of silver by the silver generator electrodes. If identical volumes of standard and unknown are used, calculations simplify to:

$$\text{Conc U} = \frac{\text{Net time U}[U_{sec} - B_{sec}]}{\text{Net time S}[S_{sec} - B_{sec}]} \times \text{Conc S}$$

where U = unknown, B = blank, and S = standard.

The Cotlove coulometer is an example of a coulometer that uses the principle of constant-current coulometry, with amperometric detection.[2] In addition to constant-current coulometry, coulometric measurements can also be performed by the controlled-potential method. In this method the electrode potential is kept constant while the current is varied continuously to maintain the desired potential.[3] The current flowing through the cell curing the reaction is related to the concentration of chloride ions in the sample. The controlled-potential method has an increased working range and lower detection limit. However, for all practical purposes constant-current coulometry is quite adequate for chloride determinations in the clinical laboratory.

Practical Considerations

The electrodes in the coulometer should be rinsed thoroughly with distilled water between titrations and be polished with a silver polish paste every day. Amperometric electrodes require cleaning by soaking in strong alkali such as concentrated ammonia. This procedure is used when the blank reading is high or the readings are not reproducible. With continuous use, the end of the generator silver anode

Figure 8–1

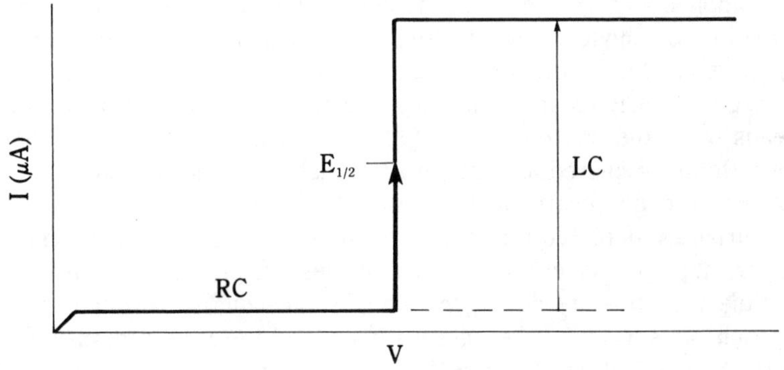

A polarogram. I = current; RC = residual current; $E_{1/2}$ = half-wave potential; LC = limiting current; V = potential.

becomes thin. The thin end should be removed and the length of the electrode adjusted to maintain it at the same level as the other electrodes.

Net titration times should be maintained between 60 to 160 seconds for optimum results. Other halogens, as well as cyanate, thiocyanate, sulfide and sulfahydryl groups interfere by scavenging the silver ions generated thereby overestimating the chloride value. Delay in separation of serum from cells can lead to chloride ion shifts into the cellular compartment, thus spuriously lowering the chloride value. Factors favoring the reformation of chloride ions from silver chloride, such as exposure to direct sunlight during titration, will lead to erroneously elevated values. The ultimate accuracy and achievement of optimum precision will depend on strict adherence to preventive maintenance schedules described in the instrument operating manual.

Polarography

Polarography is essentially a voltammetric technique in which a potential equal to the reduction potential of the element being measured is applied to a microelectrode and the resulting flow of current is measured. A graph that plots current-voltage curves is called a polarogram (Figure 8–1). Polarograms are obtained with a microelectrode (cathode), which is maintained at a potential corresponding to the reduction potential of the electroactive species to be measured and a reference electrode. Measurements are performed without stirring to avoid convection. To enable the electroactive species to reach the microcathode by diffusion only, the sample is analyzed in the presence of a 0.1 M to 1.0 M solution of a supporting electrolyte such as potassium chloride or an appropriate buffer to maintain pH. The supporting electrolyte may also contain a complexing agent, such as citrate or tartarate, to remove undesirable interfering species. The presence of the supporting electrolyte decreases the resistance of the solution and eliminates electrostatic attraction of the desired electroactive species to the microcathode under the influence of the applied voltage. The result is the movement of the electroactive species towards the microcathode solely by diffusion. A small amount of current is produced by the supporting electrolyte which is called the residual current (Figure 8–1). As the electroactive species reaches the microelectrode, the current begins to increase and plateaus when the concentration of electroactive species at the surface of the electrode becomes depleted as a result of reduction at the cathode. At this point, the surface of the microcathode is saturated with the reduced species, and the electrode surface is depleted for the remaining electroactive species in the sample by the layer (depletion layer) of reduced species around the electrode. Thus, the remaining electroactive species in the sample must diffuse through the depletion layer to reach the surface of the microcathode, and the rate of diffusion is proportional to the concentration of the electroactive species in the sample.

When the current plateaus, the microelectrode is said to be polarized; that is, it assumes the potential applied to it with little change in current. The difference between the residual current and the current at the plateau is the limiting current, which is proportional to the concentration of the desired electroactive species in the sample. The steep increase in current coincident with the reduction of the electroactive species characterizes a polarographic wave or step. The potential equivalent to half of the limiting current is called the half-wave potential ($E_{1/2}$) and is characteristic of the oxidation-reduction potential of the electroactive species. The half-wave potential can thus be used to identify an element since most elements have characteristic oxidation-reduction potentials.

The classic dropping mercury electrode (DME) is the basis of a polarographic measurement device introduced over 60 years ago. In this device, mercury from a glass reservoir is forced through a glass capillary. Minute drops of mercury are thus formed at the orifice of the capillary and fall at a constant rate. Each falling drop provides a renewable microelectrode surface at which reduction of the electroactive species occurs. The pool of mercury created by the falling drop serves as the reference electrode. Polarograms produced with DME are highly reproducible, since each falling drop provides a fresh, uncontaminated microelectrode surface. Because of the extremely small surface area of the DME, sample volumes as low as 0.01 mL can be used. The DME can be used to make measurements in the applied potential voltage range of $+0.3$ to -2.8 V versus the saturated calomel electrode. At potentials greater than $+0.3$ V, the DME cannot be used since mercury dissolves at the anode. At potentials of less than -2.8 V, interferences, such as reduction of supporting electrolyte or evolution of hydrogen, limit the usefulness of DME. The DME has largely been supplanted by newer solid electrodes, such as platinum, carbon, or wax-impregnated graphite, some of which permit measurement at positive potentials beyond the range of DME.[3]

Polarography has been widely used in the clinical chemistry laboratory for measurement of partial pressure of oxygen in whole blood (PO_2) and for monitoring enzymatic oxygen consuming reactions such as the uptake of oxygen by glucose in the presence of the enzyme glucose oxidase. The electrode devised by Dr. Leonard Clark is the basis of oxygen measurement by polarography. Current versions of the Clark electrode consist of a thin platinum wire of approximately 20 μm in

Figure 8–2

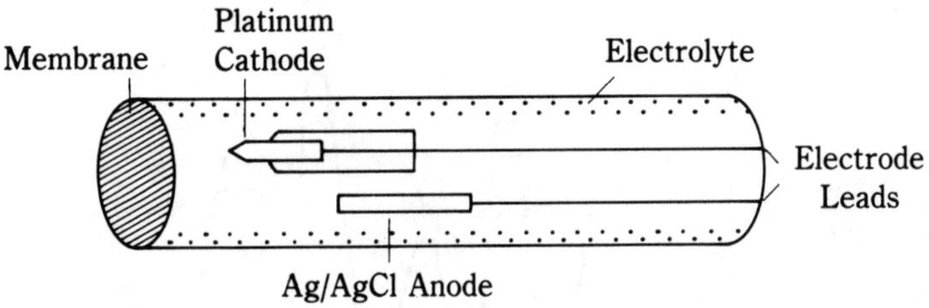

A PO_2 electrode.

diameter encased in a glass envelope with a portion of the platinum wire exposed that serves as the cathode (Figure 8–2).

The anode consists of a silver/silver chloride (Ag/AgCl) electrode. Electrical contact between the electrodes is provided by a phosphate buffer solution containing potassium chloride. A semipermeable polyethylene or polypropylene membrane allows gaseous molecules in the sample to diffuse to the surface of the microcathode. A potential of −0.6 to −0.7 V is applied between the platinum cathode and the Ag/AgCl anode. At this potential, oxygen diffusing to the surface of the microcathode is reduced, and the resulting current is proportional to the concentration or partial pressure of oxygen (PO_2) in the sample.[4]

The reactions that occur at the electrode and by dissociation of buffer ions and electrolyte are summarized:

1. Buffer ions and electrolyte dissociation

$$H_2PO_4^- \rightleftharpoons HPO_4^= + H^+$$
$$KCl \rightleftarrows K^+ + Cl^-$$

2. At the Ag/AgCl anode

$$4\ Ag^0 \rightarrow 4\ Ag^+ + 4e^-$$
$$4\ Ag^+ + 4Cl^- \rightarrow 4\ AgCl$$

3. At the platinum cathode

$$O_2 + 4e^- \rightarrow 2O^=$$
$$2O^= + 2H_2O \rightarrow 4OH^-$$
$$OH^- + KCl \rightarrow KOH + Cl^-$$

Only small amounts of oxygen are consumed because of the very small surface available at the microcathode. Consequently, current flow is approximately 1×10^{-11} A/mmHg PO_2 at 37°C. Thus, current-measuring devices should have the capability to measure picoamperes of current.

PO_2 measurements in whole blood samples are taken after calibrating gases with zero PO_2 and with a known target PO_2 value. The PO_2 of blood is dependent on temperature with values generally dropping 6% to 7% per degree centigrade

Figure 8–3

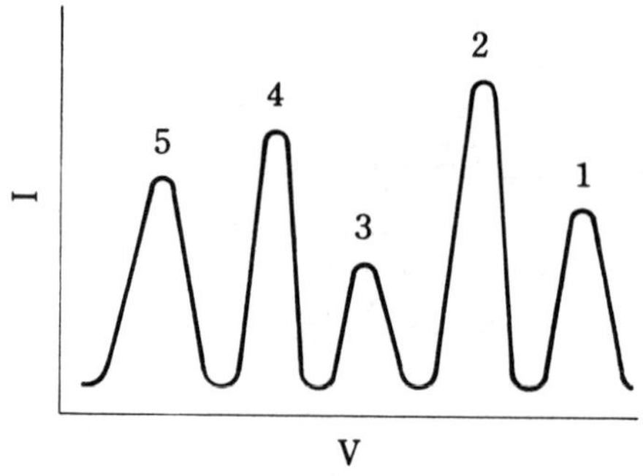

Anodic stripping voltammogram. Numbers 1 through 5 represent different elements that have been stripped at their characteristic oxidation potentials. I = current; V = potential.

(°C) decrease in temperature. Instruments for the measurement of whole blood PO_2 are adequately thermostated to bring the temperature of all samples to 37°C prior to measurement. Thus, a malfunctioning thermostat can affect the results. The platinum cathode may become contaminated with silver ions as a result of the oxidation at the anode that is brought to the cathode surface by the electrolyte. This would increase the effective surface area of the electrode, thus giving rise to high PO_2 values. The platinum cathode should be cleaned of any residual silver deposits during routine replacement of membrane, or whenever there is a drift toward high PO_2 values. Other recommended preventive maintenance procedures suggested in the instrument manual should be followed for optimum results.

Anodic Stripping Voltammetry

Essentially the technique of anodic stripping voltammetry entails the deposition or plating of the metal at the surface of a mercury film on a wax-impregnated graphite electrode by applying a negative potential for a fixed time (from 1 to 15 minutes and sometimes as long as 1 hour). The solution is stirred to ensure uniform and reproducible plating and to reduce the level of oxygen that would otherwise interfere. Once the metals are reduced at the surface of the electrode at their characteristic reduction potential, the electrode becomes plated with the reduced species.

At the conclusion of this preelectrolysis step, the electrode potential is varied linearly in the positive (anodic) direction, resulting in dissolution or stripping of the metals from the electrode surface at an applied potential characteristic of the oxidation potential of the stripping metal (Figure 8–3). The stripping or dissolution of the metal results in a current peak, the height of which is proportional to the concentration of metallic species in the sample. Peak heights can be related to the concentrations of known standards carried through the same steps as the sample. Because each element has a characteristic oxidation potential at which it is stripped during the anodic scan, the technique permits qualitative and quantitative analysis of metals in the sample.[5]

Instead of linearly increasing voltage during the stripping steps, differential pulse techniques can be used to increase sensitivity and eliminate current background. The time for the plating step can range from 30 seconds to 5 minutes with the differential pulse techniques.

Anodic stripping voltammetry permits measurement of nanogram quantities of metal and is being utilized in the simultaneous multielement analysis of elements contributing to environmental pollution, as well as elements of toxicological and nutritional significance in blood samples.

References

1. Bender GT: Clinical Instrumentation; A Laboratory Manual Based On Clinical Chemistry. Philadelphia, Saunders, 1972, pp 218–227.
2. Cotlove E: Chloride. In Seligson D (ed): Standard Methods of Clinical Chemistry, vol 3. New York, Academic Press, 1961, pp 81–92.
3. Willard HH, Merritt LL Jr, Dean TA, Settle FA. Instrumental Methods of Analysis, ed 6. New York, Van Nostrand, 1981.
4. Gambino S: Oxygen, partial pressure (PO_2) electrode method. In MacDonald P (ed): Standard Methods of Clinical Chemistry, vol 6. New York, Academic Press, 1970, pp 171–182.
5. Barendrecht E: Stripping voltammetry. In Bard AJ (ed): Electroanalytical Chemistry, vol 2. New York, Marcel Dekker, 1967, pp 53–109.

9 Chromatography

Chromatography can be traced back to the work of Tswett, a Russian botanist, who coined the term in 1906 when using the technique to separate colored pigments from leaf extracts. Martin and Synge were pioneers in the area of partition chromatography in the 1940s and were awarded the Nobel Prize in 1952 for their achievements.

Separation in chromatographic techniques is dependent on the ability of constituents in a sample to distribute themselves between two immiscible phases (stationary and mobile) in contact with each other. The stationary phase may be a solid, a liquid, or a liquid coated on a solid phase. The mobile phase, which passes over or through the stationary phase, can be either a liquid or a gas. The separation of the various constituents in a sample by chromatography will depend upon their relative affinity to the stationary phase. The extent to which a substance will distribute between the stationary and mobile phases is predicted by its distribution or partition coefficient (K), which is the ratio of the concentration in the stationary phase (CS) to that in the mobile phase (CM):$K = CS/CM$. If a substance has a K value of 1, the material will distribute equally between the two phases. A substance with a K value of less than 1 will have a lesser affinity for the stationary phase when compared with a substance with a K value greater than 1.

Different types of chromatographic techniques are possible based on the type of interaction with the stationary phase. In adsorption chromatography, the constituents in a sample are separated based on the extent of their interaction with the active chemical sites of an adsorbent used as a stationary phase. In partition chromatography, the constituents in a sample are separated on the basis of their ability to distribute themselves between the stationary and mobile phases, which in turn, will depend upon K. This partition could take place between an immobilized liquid stationary phase and a mobile phase, which may be either a liquid (liquid chromatography) or a gas (gas-liquid chromatography). In gas-solid chromatography

Figure 9–1

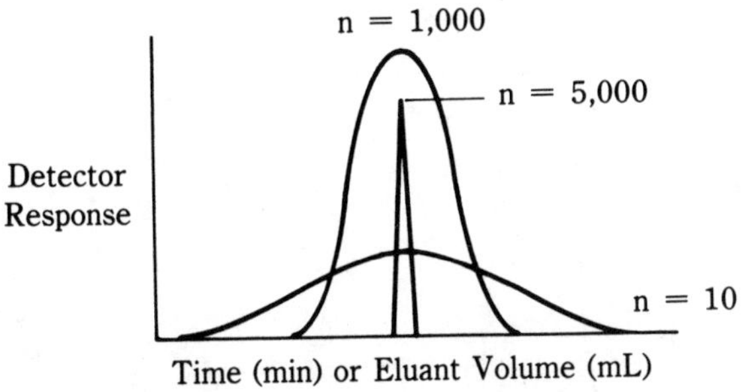

Column efficiency illustrating the effect of number of plates (n) on peak shape.

the stationary phase is a solid. If the stationary phase used is an ion-exchange resin for separating constituents based on the number and types of charges, the technique is referred to as ion-exchange chromatography. The use of a stationary phase with a controlled pore size permits separation of molecules based on their molecular size and is the basis of exclusion (or gel permeation) chromatography. The affinity of a molecule for specific chemical groups on the stationary phase is the basis of affinity chromatography.

Theoretical Plates

Ideally, to separate the many constituents in a sample, each of the constituents must have different affinities (K values) for the stationary phase to be eluted at different times (retention time, or t_r), with the mobile phase allowing good separation. To examine this theory, visualize a chromatographic column packed with a stationary phase that has been equilibrated with a mobile phase (eluting solvent or buffer). If a sample is applied to such a column and subsequently eluted with the mobile phase, the order of elution from the column will depend on the relative affinities of the constituents in the sample to the stationary phase. The shape of the eluting peaks will depend on the number of equilibrations that the constituents in a sample undergo between the stationary and mobile phase. Each such equilibration is referred to as a "plate." (The equilibration is analogous to a distillation plate used to characterize the efficiency of a distillation process.) The efficiency of a chromatographic column is characterized by the number of equilibrations or theoretical plates (n) that are dependent on the stationary phase and on the length of the column. As the value of n increases, the column becomes more efficient, producing narrower peaks. Figure 9–1 illustrates the effect of the number of theoretical plates on peak resolution.

Figure 9–2

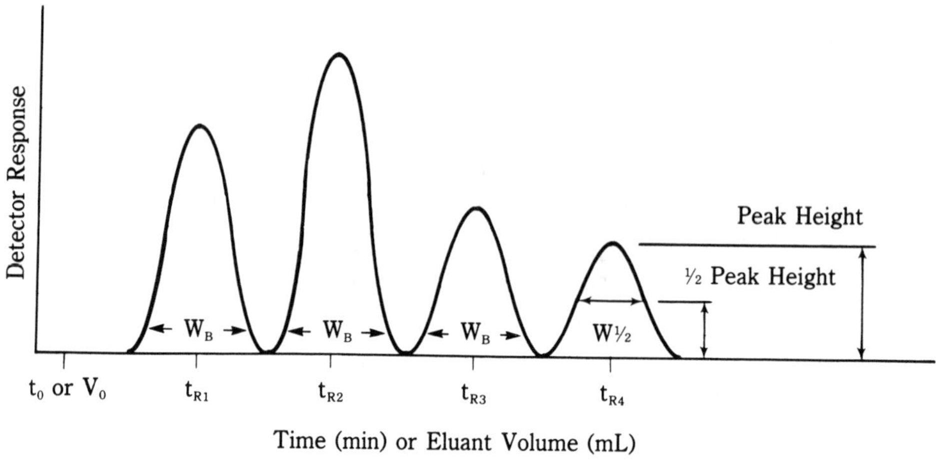

Chromatographic elution profile.

Capacity Factor

Figure 9–2 illustrates the chromatographic elution profile of sample constituents. The time required for the mobile phase or a nonretained constituent to emerge from the column is t_0, and the volume of this mobile phase is V_0 or void volume. The time required to elute each constituent is referred to as retention time (t_r), which represents the interval from sample introduction to the emergence of the midpoint of the eluant peak (T_{r1}, t_{r2}, etc). The volume required to elute a constituent is called the retention volume (V_r) and is a product of retention time and flow rate of the mobile phase (F, measured in mL/min); thus, $V_r = t_rF$. The retention of a constituent will depend on K.

A practical assessment of retention can be made by the use of the capacity factor (K′). The capacity factor is actually the effective distribution coefficient, which is obtained by multiplying K (which is the ratio of the total amount present in the stationary phase (CS) to that in the mobile phase (CM)) by the ratio of the volumes of the two phases:

$$K' = \left(\frac{\text{Total amount (CS)}}{\text{Total amount (CM)}}\right)\left(\frac{\text{Volume (CS)}}{\text{Volume (CM)}}\right)$$

Because the volume of mobile phase required to elute a constituent (V_r) can be related to the retention time, and the volume of mobile phase in the column (V_m) can be related to t_0, K′ can be calculated from the chromatogram by

$$K' = \frac{t_r - t_0}{t_0}$$

Extremely small values for K′ would imply that the constituent will be eluted close to the unretained peak, leading to poor separation. In contrast, high values for K′ would indicate that the constituent has a high affinity for the stationary phase and will elute from the column after a considerable time, resulting in broad peaks (Figure 9–2) unless columns with an unusually large number of plates are used. Optimum separation is achieved when using K′ values between 1.5 and 4.0.[1]

Column Efficiency and Resolution

The elution profile in Figure 9–2 allows assessment of the efficiency of the column since the number of plates (n) can readily be calculated using the relationship:

$$n = \left(\frac{t_r}{\sigma}\right)^2 = 16\left(\frac{t_r}{W_B}\right)^2$$

where t_r = retention time of a peak, σ = standard deviation of a peak, and W_B, the base width of the peak = 4σ. W_B is obtained by drawing tangents to the base line on each side of the inflection points of the peak. Determination of column efficiency allows an estimate of the extent to which peak broadening can occur as chromatography proceeds since high efficiency columns yield narrow peaks.

Column efficiency depends on the stationary phase used, the flow rate of the mobile phase, and the length of the column. To permit comparison of the efficiency of columns with varying lengths, it is useful to express column efficiency in terms of the height equivalent to a theoretical plate (HETP). HETP, or H, is obtained by dividing the length of the column (L) by the number of plates (n). Thus, $H = L/n$.

Ideally, the chromatographic process should result in the complete separation or resolution of the desired constituent peaks in a sample. In Figure 9–2, the resolution (R) between peaks at t_{r1} and t_{r2} can be calculated by taking the difference between their retention times ($t_{r2} - t_{r1}$) and dividing this value by the average of the widths of the two peaks; thus,

$$R = \frac{(t_{r2} - t_{r1})}{(W_{B2} + W_{B1})^2} = \frac{2(t_{r2} - t_{r1})}{W_{B2} + W_{B1}}$$

Causes of Peak Broadening

Although the time required to elute a constituent in a sample is dependent on K, not all molecules of a constituent move down the column at the same rate. Thus, the gaussian nature of the peak is caused by the peak broadening that occurs from the unequal movement of molecules through the column. The spreading or dispersion of the molecules of a constituent has many causes. For example, because of differences in column packing, some of the molecules of a constituent move faster with the mobile phase. The traverse between two stationary phase particles that are wide apart and emerge ahead of the other molecules, which traverse through

Figure 9–3

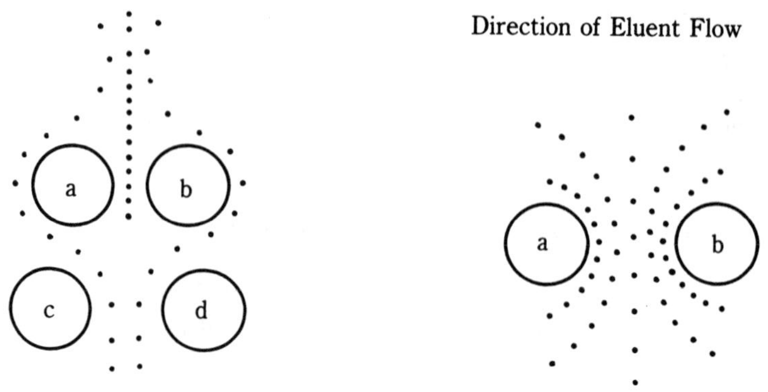

Causes of molecular dispersion. Left: Eddy diffusion—the flow of molecules (dots) between widely spaced stationary phase particles (a,b,c,d) is faster than that between more closely spaced particles. Right: Mobile phase mass transfer—molecules midway between particles move faster than those adjacent to the particles.

closely packed column-packing particles. This phenomenon is called "eddy diffusion" and is one of the factors responsible for peak broadening (Figure 9–3). Molecular dispersion also arises as a result of differential flow rate of mobile phase, with the eluant adjacent to the column packing moving slowly as compared to the eluant in the midpoint between two particles, which moves rapidly. This phenomenon called "mobile-phase mass transfer" also contributes to peak broadening.

Other causes of peak broadening are related to the time spent by the molecules as they penetrate deep into the stationary phase (stationary-phase mass transfer), the random diffusion of molecules (longitudinal diffusion), and, in porous stationary phases, the trapping of the molecules as some of the mobile phase penetrates the porous particle (stagnant mobile-phase mass transfer).

Column Selectivity

R values below 1 lead to poor separation. A 2% overlap between two adjacent peaks occurs when R = 1. When R = 1.5, two adjacent peaks are well resolved and do not overlap. Resolution between two peaks depends on how narrow the two peaks are, as judged by the base width, and how far the two peak maxima are situated. The efficiency of the column (the number of plates) determines the base width of the peak. The distance between peak maxima is related to the distribution of the various constituents between the stationary and mobile phases and is a function of column selectivity that is dependent on the proper choice of stationary and mobile phases. Column selectivity or separation between two peaks can be

Figure 9–4

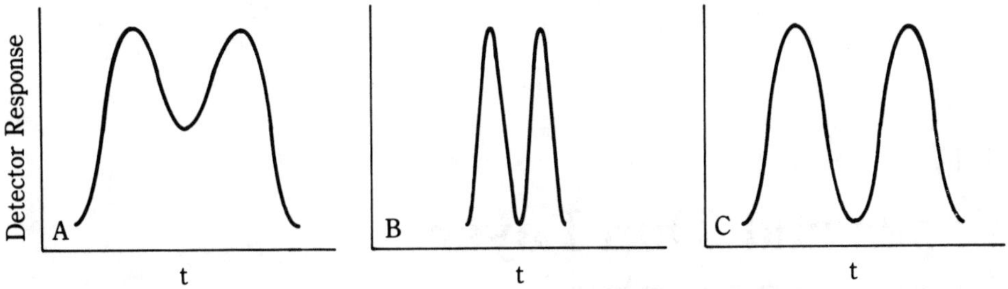

Approaches to separating overlapping peaks. A: R = 1; B: R = 1.5 by changing column efficiency; C: a change in column selectivity. t = time.

assessed by use of a selectivity factor called α. In Figure 9–2, for two peaks with retention times t_{r1} and t_{r2}, α will be given by the equation:

$$\alpha = \frac{t_{r2} - t_0}{t_{r1} - t_0} = \frac{K'_2}{K'_1}$$

In other words, α is a measure of relative retention since it is a ratio of the capacity factors (K′) of two adjacent peaks. For adequate separation, α should exceed 1 and be preferably in the range 1.05 to 2.0.

In practical terms, how can one resolve two constituents that are not well separated in a chromatographic analysis? Figure 9–4A shows two adjacent, but not fully separated peaks (R = 1). These two peaks can be completely separated either by choosing a column with a larger number of theoretical plates, increasing column efficiency (Figure 9–4B), or by varying the stationary or mobile phase and favorably changing the column selectivity factor (Figure 9–4C). In summary, an appreciation of the theoretical aspects of chromatography better equips the laboratorian to solve practical analytical problems.

Reference

1. Giddings JC: Dynamics of Chromatography, Part 1: Principles and Theory. New York, Marcel Dekker, 1965.

10
Paper and Thin Layer Chromatography

Paper Chromatography

In paper chromatography, which is a form of partition chromatography, partition of the constituents in a sample occurs between a mobile solvent phase and a stationary phase where water is hydrogen-bonded to cellulose (paper).[1] Separation depends on the rate of diffusion, the nature of the solvent mobile phase, and the solubility of the constituents in the sample. The amount of water within the cellulose meshwork of paper determines the degree of separation achieved. The water content of paper is usually 20% by weight.

The porosity of paper, coarse, medium, or fine, determines the rate of migration of the developing solvent across the paper surface. The rate of movement of the solvent is fastest with the coarse-porosity paper and slowest with low-porosity paper, although low-porosity paper provides the best resolution of the separated components in the sample. The mobile solvent phase is usually dependent on a solvent or a mixture of solvents that are generally miscible with water. In some cases, water is incorporated as a component of the solvent mixture. Even though solvents miscible with water are used, they are immiscible with water bonded to the stationary phase, thus meeting the conditions required for partition between two immiscible phases.

Spotting

Under conditions of controlled humidity, minute drops of the sample are applied to the paper using either a capillary pipette or a precision syringe. The initial drop of sample in a volatile solvent is allowed to dry, preferably with the aid of a hair dryer, before the next drop is applied. Sample application is performed in an area

Table 10–1 Typical Solvent Systems Used in Paper Chromatography and Their Applications

Solvent System (vol/vol)	Application
n-Butanol/acetic acid/water (40:10:50)	Amino acids
n-Butanol/pyridine/water (50:28:22)	Simple sugars

approximately 2 cm from one end of the paper. For optimum separation, the diameter of the sample spot should not exceed 2 mm, and the concentration of each drop should be in the range of 10 to 50 μg. If the total sample concentration exceeds 500 μg, and the proper spot diameter is not achieved, band spreading and poor separation will occur. Thick papers designed for applying an increased amount of sample should be used when a large quantity of the desired constituent must be separated.

Development

After the sample is spotted, the paper is developed in an appropriate solvent system. The solvent system selected depends on the chemical nature of the constituents to be separated. Table 10–1 lists common solvent systems used in paper chromatography with their applications. Inclusion of an acid (eg, acetic acid) or a base (eg, pyridine) in the butanol-water system allows separation of the polar sugars and amino acids. A limitation in multisolvent systems is the formation of multiple bands due to the partition of the individual solvents with the paper, which may interfere with the quality of separation of sample constituents.

The development process is accomplished by using either the ascending or the descending flow of the mobile phase solvent (Figure 10–1).

The paper is developed in an air-tight glass chamber. Prior to development, the chamber is saturated with solvent vapor for 1 to 3 hours by placing solvent at the base of the chamber. A filter paper is wrapped around the inside of the chamber to facilitate equilibration with solvent vapor. The solvent rises by capillary action and moistens the filter paper, thus permeating the inside of the developing chamber with solvent vapor. The equilibration of the chamber prior to development minimizes the phenomenon of band streaking and also allows the migration distances of the spots to be reproducible from run to run. In the ascending development procedure, the paper is introduced into the equilibrated chamber so that the sample application spots on the paper are slightly above the level of the developing solvent at the bottom of the chamber. With the lid in place, the solvent will begin to migrate up the paper by capillary action; the partition effects produced between the solvent and paper will result in separation of the constituents. In the descending development procedure, the developing solvent is placed in the trough just below the lid of the chamber. With a glass rod, the paper is suspended from the trough with the sample spot close to, but not actually dipping in, the solvent. Care should be taken to ensure that the base of the paper is not dipping in the equilibration solvent at the bottom of the chamber. As the solvent migrates down the paper by capillary action and by gravity in the descending technique, a more rapid flow is achieved

Figure 10–1

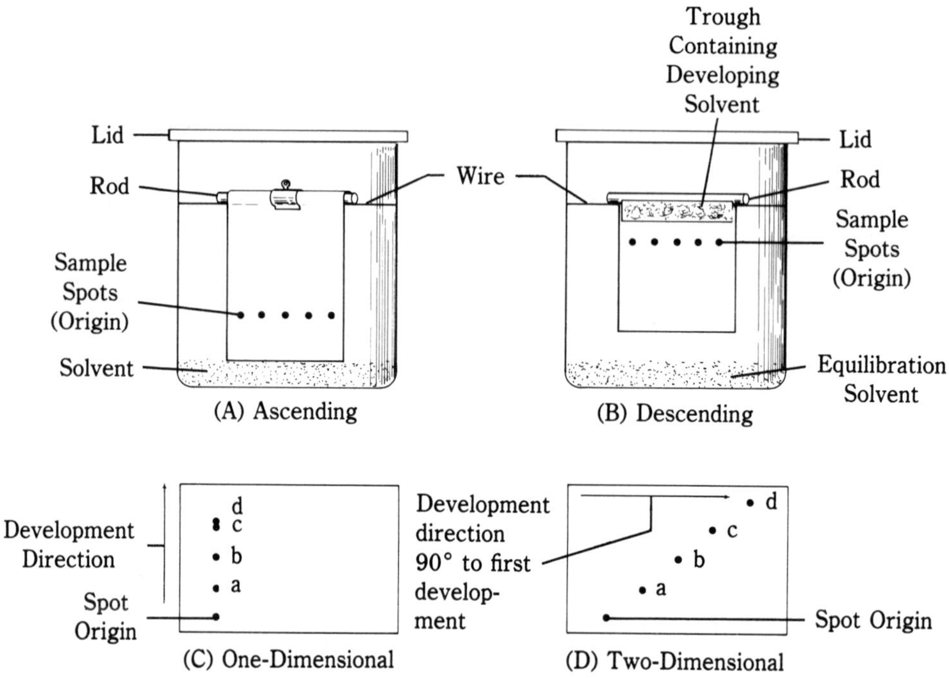

Chromatography development (A and B). (C,D) One-dimensional vs two-dimensional development: a, b, c, and d are isolated constituents. Constituents c and d, unresolved in one-dimensional development (C), are resolved in two-dimensional development (D).

as compared with the flow in the ascending technique; however, the ascending technique is popular because of its simplicity.

Ideally, the solubility of the sample constituents in the developing solvent should be minimal since increased solubility would lead to migration with the mobile phase. This would result in poorly resolved chromatograms because of the concentration of separated bands close to the solvent front.

In a one-dimensional solvent system (Figure 10–1C), the development can take 2 to 4 hours depending on the paper used. Following development, the paper is removed, and the solvent front is marked with a pencil and allowed to dry until the separated spots are visible. Alternatively, the dried paper could be developed in a second dimension if it is turned 90° and developed in a second solvent (two-dimensional chromatography, Figure 10–1D). If proper solvents are chosen for use in the second dimension, constituents that were unresolved in the one-dimensional procedure can be separated (Figure 10–1D). In the second dimension, separation is improved by increasing the distance of migration. The paper used is usually 20 cm × 20 cm (W,L) as compared with the 5 cm × 20 cm (W,L) paper used in the one-dimensional solvent system.

Visualization

If the constituent spots do not have visible color on the chromatography paper, appropriate reagents can be sprayed to allow visualization. For example, when chromatography paper containing amino acids is sprayed with a 0.2% solution of ninhydrin, a pale blue-violet color develops. An ultraviolet lamp allows constituents that are either fluorescent compounds, or that absorb in the ultraviolet region, to be detected. Radioactive compounds can be detected either with a radiochromatogram scanner or by autoradiography, where they will appear as dark areas on an x-ray film. Spraying with 50% sulfuric acid to char and thereby make the constituents visible as brown spots is not recommended since the chromatography paper will be destroyed by sulfuric acid.

Identification

Each constituent has a certain R_F value that is constant under defined conditions and is dependent on the partition coefficient of the constituent. The R_F value is defined as the ratio of the distance moved by the constituent (measured from the origin to the center of the spot) to the distance moved by the solvent front (Figure 10–2), where

$$R_F = \frac{\text{Distance moved by the constituent}}{\text{Distance moved by the solvent front}}$$

In general, R_F values of 0.4 to 0.8 are adequate for providing good resolution. Identification can also be related to the migration rate of a substance of the same chemical class. Thus, migration of carbohydrates can be related to the migration rate of glucose (R_G values), where

$$R_G = \frac{\text{Distance moved by a carbohydrate}}{\text{Distance moved by glucose}}$$

Relative migration rates are particularly helpful when the solvent is allowed to run off the paper, as is done sometimes in descending chromatography to increase the migration distances of constituents in a sample.

Applications

One of the earliest applications of paper chromatography was in the identification of characteristic amino acids excreted in aminoacidurias. The technique is also useful for the identification of various sugars excreted in melliturias. The advent of rapid thin layer chromatographic procedures has diminished the routine use of paper chromatography.

Figure 10–2

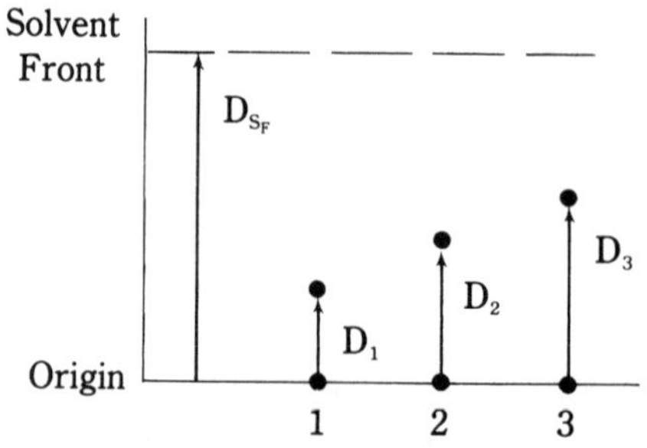

Determining the R_F value. D_1, D_2, and D_3 are the migration distances of constituents 1, 2, and 3 respectively from the origin to the center of the migrated spot. D_{S_F} is the distance from the origin (——) to the solvent front (— — —). Hence:

$$R_{F_1} = \frac{D_1}{D_{SF}}; \; R_{F_2} = \frac{D_2}{D_{SF}} \text{ and } R_{F_3} = \frac{D_3}{D_{SF}}$$

Figure 10–3

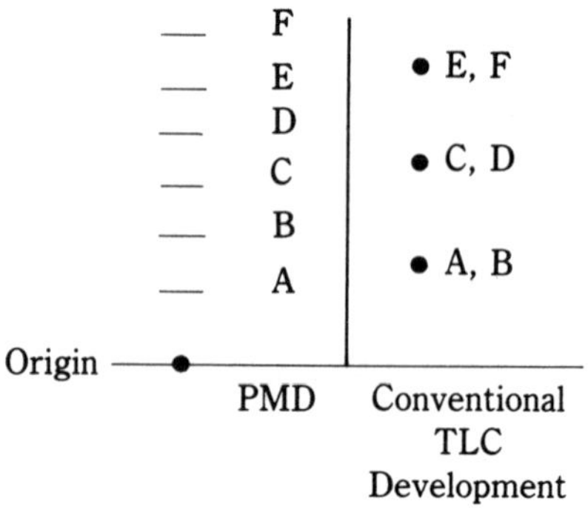

Resolution by PMD compared with conventional TLC Development. Two constituents with very close R_F values are resolved by PMD as narrow bands in contrast to conventional TLC development.

Thin Layer Chromatography

In the technique of thin layer chromatography (TLC), a thin layer of adsorbent approximately 0.1 to 0.3 mm in thickness (for analytical determinations) and approximately 2 to 10 mm (for preparative purposes) is mixed with a binder with the resulting slurry deposited on a glass or plastic plate.[2] Applicators are available for spreading slurry. The adsorbent commonly used is silica gel, although aluminum oxide (alumina) and diatomaceous earth (kieselguhr or celite) have also been used. The binder commonly used is calcium sulfate, which enables the adsorbent to stick firmly to the glass or plastic plate that serves as a backing.

The amount of binder used is approximately 5% to 10% of the weight of the adsorbent. Glass fiber sheets can be used instead of conventional thin layer plates and are advantageous because they allow quicker separation. An adsorbent such as silicic acid can be impregnated on the glass fiber sheets.

The recent availability of glass plates precoated with silica gel having an extremely narrow particle size distribution (between 2 and 5 μm) and controlled pore diameter (approximately 6 nm) has improved detection limits ten times those previously possible in TLC. For the high-performance TLC plates, sample requirements are one tenth the amount needed for use with the conventional TLC plates.[3]

In addition to adsorption chromatography, ion exchange or gel permeation chromatography can be performed on thin layer plates by the appropriate choice of media: ion exchange celluloses for ion exchange chromatography and dextran for gel permeation chromatography.

The steps described for paper chromatography such as equilibration of the chamber, spotting, development, and visualization are, in general, also applicable to TLC.

Spotting

Before the chromatography plates are spotted, they should be "activated" to remove any water that may affect the efficiency of separation. This drying procedure is done by placing the plates (held in aluminum racks) in an oven at 100 to 120°C for approximately one-half hour. The activated plates are stored in a cabinet that contains a dessicant, such as an indicating silica gel.

In conventional TLC 20 × 20 cm plates are used, with the samples spotted 2 cm from one end. The diameter of the sample spots should be 3 to 6 mm, and the sample size should be between 1 and 5 μL (10 to 100 μg).

As in paper chromatography, the initial sample drop should be dried with a hair dryer or with a stream of nitrogen before the next drop of sample is applied. With high-performance TLC plates (10 × 10 cm), sample size can be as small as 0.1 to 0.2 μL with the sample spot having an approximate diameter of 1.0 mm.

Development

The spotted plates are developed in a glass chamber saturated with vapor of the developing solvent in the same manner as described for paper chromatography. Development can be in either a one- or two-dimensional solvent system. Development techniques are available to improve separation[4] and can be broadly classified into five categories: continuous, multiple, programmed multiple, stepwise, and gradient elution.

In continuous development, the solvent is continuously introduced over the edge of the plate where it evaporates. In multiple development, the TLC plate is developed once, dried, and redeveloped one or more times with the same solvent. In programmed multiple development (PMD), the developing solvent is allowed to rise on the plate for a preset period of 10 to 100 seconds. The solvent flow is interrupted, and the solvent is evaporated by heat or by a flow of inert gas for a preset period. At the end of this drying cycle, the solvent is allowed to rise again for a longer period; the solvent flow is interrupted, and the drying cycle is recommenced. A number of such solvent rise and drying cycles constitute PMD.[5] As compared with conventional TLC, where development time is fixed, there are many development cycles in PMD; the word "programmed" refers to the increase of the development time with each development cycle. PMD is advantageous because of the successive solvent rise and drying cycles, which have the effect of concentrating the developed spots that appear as narrow bands, as opposed to the rounded spots that appear in conventional development (Figure 10–3). In addition, PMD allows constituents with very close R_F values to be resolved. In stepwise development, the TLC plate is developed in one solvent, dried, and developed again in another solvent, preferably in the second dimension. In gradient elution development, the solvent composition is continuously varied.

Further improvements in separation can be achieved by spotting on a TLC plate with two adjacent layers of different composition, such as silica gel constituting one layer, and silica gel impregnated with silver nitrate on the other layer. Constituents unresolved on the silica-gel layer may be resolved by chemical reaction with silver nitrate in the adjacent layer. Another approach is to develop the TLC plate in one dimension, effect a chemical reaction by spraying, and redevelop in the second dimension. This procedure is the basis of the separation-reaction-separation (SRS) technique.

Adsorbents and Solvents

Silica gel is the all-purpose adsorbent used in TLC. Diatomaceous earths such as kieselguhr have been used for separation of carbohydrates, lipids, and amino acids, while celite has been used for separation of steroids. The choice of the developing solvent is dependent on the type of chromatographic procedure involved. In adsorption chromatography, neutral solvents such as hexane/ether/acetic acid (90:10:1 [vol/vol]) or chloroform/methanol (95:5 [vol/vol]) are used. Resolution is on the basis of chemical nature (type) and the number of functional groups. Where partition effects are involved, solvents such as butanol/acetic acid and water are applicable as described in paper chromatography.

Visualization and Quantitation

A variety of visualization techniques are available. The TLC plate may be precoated with fluorescent indicator to visualize constituent bands by fluorescence quenching under ultraviolet light illumination at 254 nm.

Spraying with fluorescamine has permitted detection and quantitation by fluorescent densitometry of nanogram levels of amino acids and of drugs (ie, amphetamine and its analogs) that form fluorescent derivatives with this reagent.

Since numerous spraying agents are available for visualization of drugs, TLC has broad application in the clinical laboratory. For example, diphenylcarbazone is used for visualizing barbiturates, and iodoplatinate is used for differentiating meprobamate from barbiturates, since barbiturates do not react. To detect acidic (barbiturates), neutral (glutethimide), and basic (morphine) drugs in a sample, a series of sprays could be used in the following order: diphenylcarbazone, mercuric sulfate-sulfuric acid, iodoplatinate, and bismuth subnitrate in glacial acetic acid (Dragendorff solution).[6]

Spraying agents for benzodiazepine drugs include iodoplatinate, dichloroquinonechloromide, and Bratton-Marshall reagent (0.4% N-α-naphthylene diamine hydrochloride in acetone/water (4:1 [vol/vol]).[7]

Other general-purpose sprays include iodine vapor for visualizing unsaturated compounds and antimony chloride ($SbCl_3$) in glacial acetic acid for visualizing steroids.

Applications

In addition to detection and quantitation of drugs,[8] TLC is utilized in the detection of lipids, steroids, carbohydrates, nucleotides, and amino acids. Recently, procedures have been developed for the visualization of phospholipids in amniotic fluid, which serve as an indicator of fetal lung maturity.[9–11]

Thin layer chromatography offers a rapid identification technique with development time ranging from 10 to 100 minutes depending on the separation required. Quantitation can be done either by densitometry or by having the isolated spots of constituents eluted and measured by colorimetric and gas-liquid chromatographic procedures. Recently, the various steps of TLC have been automated.

In one such system, 34 samples for amino acid analysis were processed in 1 hour, in comparison with the manual technique, which required a 40-minute development at 20°C, and 7.5 minutes of drying time at 90°C per sample.[12] Besides having increased speed, the automated systems have the advantage of attaining reproducible R_F values since the various TLC steps are accurately controlled.

In conclusion, both paper and thin layer chromatography are powerful screening and detection techniques. TLC, in particular, has proved valuable as a qualitative screening technique in the emergency toxicology laboratory with the availability of systems with drug standards incorporated in thin layer plates that permit comparison of relative migration and staining characteristics of unknown drugs to standard drugs.[13]

References

1. Hais IM, Macek K: Paper Chromatography. New York, Academic Press, 1963.
2. Stahl E: Thin Layer Chromatography: A Laboratory Handbook, ed 2. New York, Springer-Verlag, 1969.
3. Fenimore DC: High performance thin layer chromatography. Anal Chem 53:254A, 1981.
4. Mangold HK: Thin layer chromatography. In Curtius HC, Roth M (eds): Clinical Biochemistry, Principles and Methods, vol 1. New York, Walter de Gruyter, 1974, pp 13–45.
5. Jupille TH, McNair HM: Quantitative aspects of programmed multiple development. In Wham WN, Halaby KS, Morris SJ, Scott FI, Jr (eds): Laboratory Instrumentation: Chromatography. Fairfield, Conn, International Scientific Communications, 1977, series 2, vol 1, pp 334–338.
6. Davidow B, LiPetri N, Quame B: A thin layer chromatographic screening procedure for detecting drug abuse. Am J Clin Pathol 50:714, 1968.
7. Swada H, Akira H, Asano S, Matsumoto Y: Isolation and identification of benzodiazepine drugs and their metabolites in urine by use of Amberlite XAD-2 resin and thin layer chromatography. Clin Chem 22:1569, 1976.
8. Clark DR, Hajar TM: Detection and confirmation of cocaine use by chromatographic analysis for methylecgonine in urine. Clin Chem 33:118, 1987.
9. Pappas AA, Mullins RE, Gadsden RH: Improved one-dimensional thin layer chromatography of phospholipids in amniotic fluid. Clin Chem 28:209, 1982.
10. Spillman T, Cotton DB, Lynn SC, Jr, Bretaudiere J: Influence of phospholipid saturation on classical thin-layer chromatographic detection methods and its effect on amniotic fluid lecithin/sphingomyelin ratio determinations. Clin Chem 29:250, 1983.
11. Coapman-Hankin RA, Kiechle FL, Epstein E, Artiss JD, Zak B: Three methods compared for determining phosphatidyl glycerol in amniotic fluid. Clin Chem 31:1374, 1985.

12. Brandt KM: Automated thin-layer chromatography. In Wham WN, Halaby KS, Morris SJ, Scott FI, Jr (eds): Laboratory Instrumentation: Chromatography. Fairfield, Conn, International Scientific Communications, series 2, vol 1, 1977, pp 309–313.
13. Finkle HI: Phencyclidine identification by thin layer chromatography. A rapid screening procedure for emergency toxicology. Am J Clin Pathol 70:287, 1978.

11
Ion-Exchange Chromatography

Ion-exchange chromatography utilizes the attraction of ionized species in a sample to oppositely charged ionic groups on the surface of a stationary phase. The retained sample constituents are subsequently displaced by the eluting buffer in the order of their affinity (attraction) for the stationary phase, with the species having the least affinity for the stationary phase eluting first.

Stationary Phases

Although a variety of stationary phases have been utilized to attach charged functional groups, the more commonly used matrices include styrene-divinylbenzene copolymer, cellulose, dextran, and, with the advent of high-performance liquid chromatography (HPLC), bonded-surface coated or pellicular particles or bonded-porous-phase particles with a silica matrix. In the styrene-divinylbenzene–type resin, the amount of divinylbenzene used, which varies from 2% to 12%, controls the degree of cross-linking. Increasing the concentration of divinylbenzene, the cross-linking agent, reduces the pore size of the resin and also renders the resin structurally rigid and insoluble. In practice, the average concentration of divinylbenzene used is 8%. Ion-exchange groups are attached to specific sites on the resin bead. Figure 11–1 illustrates the structure of a styrene-divinylbenzene ion-exchange resin bearing ion-exchange groups.

Large molecules that have difficulty in penetrating the cross-linked structure of styrene-divinylbenzene resin can be separated either on cellulose matrices, which have a fibrous structure, or on dextran. Ion-exchange groups attached to cellulose or dextran bind proteins weakly in contrast to the strong binding associated with styrene-divinylbenzene ion-exchangers. The weakly bonded proteins are easily dissociated from the resin and are eluted without the risk of denaturation.

Figure 11–1

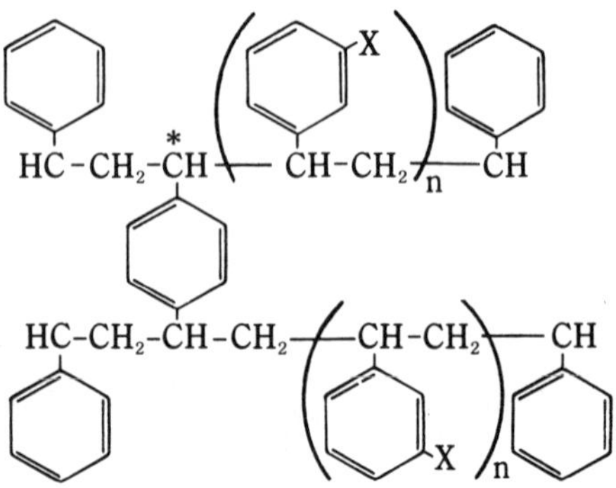

Structure of a styrene-divinylbenzene polymeric resin. X = ion exchange group; * = divinylbenzene cross-linking molecule; and ($)_n$ represents repeating units of the polymer.

Examples of stationary phases used in HPLC[1] are pellicular particles with a thin ion-exchange layer coated onto glass beads and ion-exchange groups coated or bonded to microparticles of silica, which themselves are coated onto a glass bead to form superficially porous resins. These bonded phases provide sample ions ready access to ion-exchange groups in contrast to styrene-divinylbenzene resins, in which the sample ions have to penetrate the resin matrix to reach the ion-exchange groups.

Ion-Exchange Groups

Negatively charged ion-exchange groups that are acids are used to separate positively charged ions (cations). Chemically, these cation-exchange groups can be a strong acid, such as sulfonic acid ($-SO_3H$), which is introduced by sulfonating the stationary phase resin, or the functional group can be a weak acid, such as a carboxylic acid ($-COOH$).

For separation of negatively charged ions (anions), positively charged ion-exchange groups that are strong or weak bases are also used. An example of a strong-base anion-exchange functional group is trialkylammonium chloride ($-CH_2N(CH_3)_3^+Cl^-$), which is introduced by reacting the resin with chloromethyl ether and subsequent treatment of the chloro groups with the tertiary amine, trimethylamine. Examples of weak-base anion-exchange functional groups are amines ($-NH_3^+Cl^-$) or secondary amines ($NHR_2^+Cl^-$).

Diethylaminoethyl groups (DEAE) attached to cellulose or dextran are examples of weak-base anion-exchange resins used for separation of proteins. Examples of cation-exchange functional groups attached to cellulose or dextran are carboxymethyl (CM-cellulose) and sulfoethyl (SE-cellulose).

Water-insoluble organic liquids coated on porous or pellicular resins have been used as liquid ion exchangers. Dialkyl esters of phosphoric acid (bis-(2-ethylhexyl)-phosphoric acid) and alkyl esters of alkane phosphoric acids are examples of liquid cation exchangers. High-molecular weight amines, such as trioctylamine, and quaternary ammonium compounds, such as trioctylmethyl ammonium, serve as liquid anion exchangers.

Mechanism of Ion Exchange

In cation exchange, sample cations compete with mobile-phase (buffer) counter ions for negatively charged ionic groups on the surface of the resin:

$$R^-X^+ + Y^+ \rightleftharpoons R^-Y^+ + X^+$$

where R^- represents ionic groups in the resin, X^+ are mobile-phase counter ions, and Y^+ are sample ions undergoing exchange.

The mechanism is reversed in anion exchange because sample anions compete with mobile-phase counter ions for positively charged ionic groups on the resin:

$$R^+X^- + Y^- \rightleftharpoons R^+Y^- + X^-$$

where R^+ represents ionic groups in the resin, X^- are mobile-phase counter ions, and Y^- are sample ions undergoing exchange. In general, an increase in atomic weight, ionic size, and charge results in increased retention of sample ions on the oppositely charged ion-exchange resin. An increase in the degree of hydration, however, will cause affinity of the sample ion for the resin to decrease. Retention of sample ion on ion-exchange resin will also decrease with an increase in the ionic strength of the buffer, since the buffer counter ions will compete strongly for ion-exchange sites on the resin.

Classification of Ion-Exchange Resins

Ion-exchange resins can be classified on the basis of pore size (degree of cross-linking), the particle size, the type of ion-exchange group, and the exchange capacity of the resin (ion-exchange capacity).

Pore Size

Porous polymeric resins swell in buffer in relation to their pore size. The pore size is governed by the degree of cross-linking agent used in polymerizing the resin. Highly cross-linked resins limit pore size and impede the flow of large molecules, thus preventing their participation in the ion-exchange process.

Particle Size

The core of the pellicular particles is composed of glass beads approximately 30 μm to 40 μm in diameter, which are coated with an ion-exchange resin layer 1 μm to 2 μm in thickness. The superficially porous bonded silica particles are approximately 5 to 10 μm in diameter. In contrast, the particle size of the conventional styrene-divinylbenzene resin is 7 to 10 μm.

Ion-Exchange Group

The type of ion-exchange group used governs the operating pH range for chromatography. Both strong-acid cation-exchange and strong-base anion-exchange resins have a broad operating pH range. Strong-acid cation exchangers will be charged at a pH range of 2 to 13. Below pH 1, the charge of the ionic group on the resin is neutralized by protons (H^+) and thus is ineffective. The ionic groups of strong-base anion-exchange resins are charged in the pH range 1 to 10. Above pH 11 the ionic groups on these anion-exchange resins are neutralized by base (OH^-), thus interfering with their participation in ion exchange. A weak-acid cation exchanger such as carboxylic acid (—RCOOH) will be unionized at acid pH and will not enter into ion exchange.

The operating pH of these weak-acid cation exchangers lies between pH 6 and pH 13. A practical implication is that at an acid pH, a weak base will be unable to raise the pH above 6, whereas a strong base can, which results in selective retention and separation of strong base from weak base. Similarly, a weak-base anion exchanger such as $-NH_3^+Cl^+$ will not be ionized at alkaline pH. These exchangers should be used between pH 8 and pH 1. Weak-base anion exchangers at alkaline pH (above 8) will be unionized and will not enter into ion exchange with a weak acid, whereas a strong acid, by driving the pH below 8, can, resulting in selective separation of a strong from a weak acid. In general, strong-acid and strong-base ion exchangers can separate both strong and weak ionic compounds. Weak-acid or weak-base ion exchangers are useful in separating compounds, such as proteins, that would be denatured in the process of dissociation from tightly bound strong-acid or strong-base resins.

The influence of pH on separation in ion-exchange chromatography can be exploited by proper choice of mobile-phase buffer. Increasing the pH of the mobile phase results in increased retention of sample anions to anion exchangers, but decreases the attraction of cations to cation exchangers. Conversely, decreasing pH of the mobile phase favors retention of sample cations to cation exchangers and decreases the affinity of anions to anion exchangers. A knowledge of the PK_a of acidic and basic components of the sample to be separated can help in judiciously choosing the proper mobile-phase pH. (PK_a and PK_b represent the negative logarithm of the ionization constant of an acid and base, respectively.) According to the Henderson-Hasselbach equation, when a pH is equal to the PK_a of a weak acid or weak base (PK_b), half of the sample will be in the ionized form. For a weak acid, a choice of mobile-phase pH well below its PK_a will favor the unionized form and thus will be retained weakly. For a weak base, a choice of mobile-phase pH well above its PK_b will favor the unionized form and thus affect its retention on the ion-exchange resin.

The choice of mobile-phase counter ion also determines the retention of sample to the ion-exchanger. Thus on cation-exchange resins, sample retention is stronger with sodium as the mobile-phase counter ion than with potassium. Similarly, sample retention is stronger on anion-exchange resins when the mobile-phase counter ion is sulfate, as compared with chloride or fluoride.[2]

Ion-Exchange Capacity

The number of ion-exchange groups determines the capacity of the stationary phase for exchanging sample ions. This is expressed in terms of ion-exchange capacity, which for porous styrene-divinylbenzene resins is in the range of 3 to 5 mEq of counter ions or sample ions that can be exchanged per gram of resin. The bonded-phase superficially porous silica particles have a lower ion-exchange capacity: 0.5 to 2 mEq/g. The pellicular particles have the lowest ion-exchange capacity (0.01 to 0.1 mEq/g). In general, because sample size should be limited to less than 5% of the ion-exchange capacity of the exchanger to ensure maintenance of column efficiency, pellicular resins can tolerate a sample size one-hundredth (0.05 mg/g of resin) of that used on a styrene-divinylbenzene polymer resin (50 mg/g of resin).

Preparation of Columns

Conventional styrene-divinylbenzene resins are allowed to swell completely in distilled water prior to packing in a column. The extent of swelling is dependent on the resin pore size, as is the time required for the resin to swell to the maximum size. The supernatant particles that do not settle are designated as "fines" and decanted. The suspension in distilled water and the decantation steps are repeated several times until all the fines are removed that would otherwise slow down the flow rate of buffer during the chromatographic procedure. When the resin is completely swollen, it is treated further prior to packing in a column. For an anion-exchange resin, the treatment involves washing with 0.5N sodium hydroxide to neutralize acidic impurities, a distilled-water washing step to remove residual alkali, washing with 0.5N hydrochloric acid, and finally with distilled water to remove residual acid. The resin is now in chloride form and ready for use. For a cation-exchange resin, the sequence of steps is reversed since the resin will be suspended in the sodium form. The column packed with resin slurry is equilibrated with buffer prior to applying sample and eluting with buffer.

With the advent of HPLC, a variety of efficient packing techniques have become available. In one such technique the slurry of the swollen ion-exchange resin of appropriate density adjusted with calcium chloride is packed into the column under high pressure.

After sample application, the selective elution of the various constituents in the sample can be best achieved by a gradient of increasing salt concentration at constant pH. Gradient elution can conveniently be performed using gradient mixers. Essentially, this entails using two reservoirs. One of the reservoirs, the primary-buffer reservoir, contains the eluant buffer and is connected to the column. This reservoir is continuously stirred with a magnetic stirrer during the chromatographic procedure. To this reservoir is connected another reservoir containing buffer with added salt. As chromatography proceeds, mixing buffer of higher ionic strength with the buffer in the primary reservoir gradually changes the salt concentration

Figure 11–2

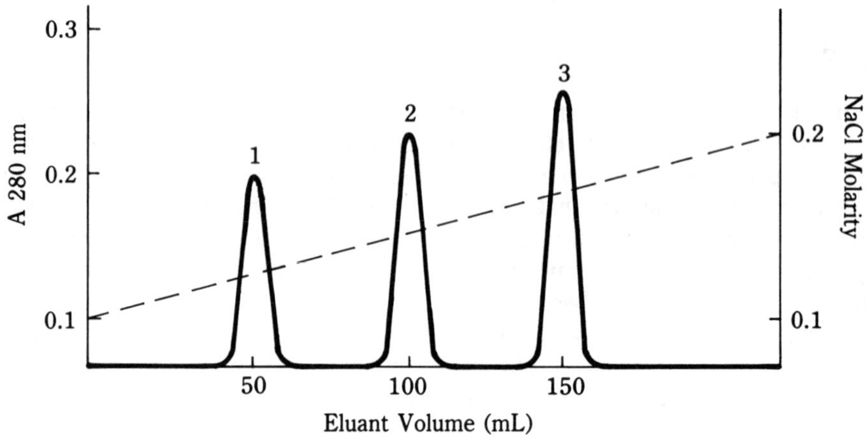

Gradient elution chromatographic profile depicting eluant volume (mL) vs absorbance (280 nm) and NaCl molarity. Elution of three proteins on an anion-exchange column. Protein 1 has the least net negative charge and is eluted first with a small increase in ionic strength. Further increases in ionic strength elute protein 2, which has a greater net negative charge than protein 1. Protein 3, which has the greatest net negative charge, is eluted last. A = absorbance.

of eluting buffer and effects displacement of retained sample constituents (Figure 11–2). The relative size of the two connecting reservoirs will determine whether the gradient will be linear, concave, or convex.

In manual ion-exchange chromatographic procedures performed in the laboratory with mini-columns, such as the determination of creatine kinase isoenzymes, displacement of the isoenzymes can be done readily by eluting the column with buffers of increasing salt concentration, which in effect results in stepwise increments in ionic strength.

Stability of Ion Exchangers

The styrene-divinylbenzene–type ion exchangers are generally stable up to 100°C with the exception of strong-base anion-exchange resins in the hydroxide form, which decompose at 60°C. Strong oxidizing and reducing agents decompose the polymeric resins and should be avoided. The bonded ion-exchange resins are stable up to 60°C except at highly alkaline pH.

Applications

The list of biological constituents that can be separated by ion-exchange chromatography is extensive, and procedures utilizing this technique are available for amino

Figure 11–3

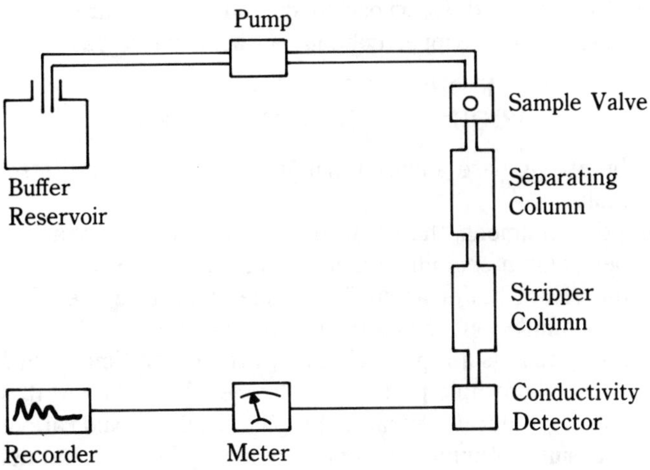

Apparatus for ion chromatography.

acid and drug analysis, purification of proteins,[3] measurement of hormones (eg, T_4), glycosylated hemoglobin,[4] isoenzymes of creatine kinase, lactate dehydrogenase, and N-acetyl-β-D-glucosaminidase,[5] and analysis of polyamine metabolites in cancer. The current availability of high-efficiency bonded-phase ion-exchange packing materials for HPLC has improved resolution and reduced the analysis time required for ion-exchange chromatographic procedures.

Until recently, ion-exchange chromatography was limited to constituents that could be measured by absorbance or fluorescence measurements. Thus, amino acids were measured by reaction with ninhydrin, proteins were measured by their absorbance at 280 nm in the ultraviolet region, and isoenzymes were measured by reacting the column effluents with their respective substrates and subsequent ultraviolet or fluorometric estimation. No procedure existed for the direct measurement of ionic constituents in a sample in the presence of the mobile-phase eluant buffer ions.

This limitation has been overcome in the two-column technique called "ion chromatography"[6] (Figure 11–3). The sample is injected into a first column through which the eluant mobile-phase buffer is pumped. Column effluent from this separating column is fed to a second ("stripper") column that neutralizes the buffer ions, leaving separated ionic constituents in the sample unaffected. The constituents are led through a conductivity cell in a background of de-ionized water. The conductivity of the ions is monitored in a meter and the conductivity response is displayed as a recorder peak, the height or area of which is proportional to the concentration of the eluting sample ion. For example, if the sample contains cations such as sodium, ammonium, and potassium as their chloride salts, they are first separated on a cation-exchange column using dilute hydrochloric acid (HCl) as the eluant.

The stripper column used in this case is a strong-base anion-exchange resin in the hydroxide form. The HCl eluant is neutralized, while the metal chlorides are converted into their hydroxides, whose conductivity is measured in the conductivity cell. The reactions in the stripper column can be visualized as follows:

$$R^+OH^- + HCl \rightarrow R^+Cl^- + H_2O$$
$$R^+OH^- + M^+Cl^- \rightarrow M^+Cl^- + R^+Cl^-$$

where R^+ is the strong-base anion-exchange resin and M^+ is the separated metal ions in the sample.

For separation of anions, the separating column consists of an anion exchanger with the stripper column containing a strong-acid cation exchanger in the H^+ form to neutralize the eluant buffer anion (hydroxide or bicarbonate). The conductivity of the remaining sample ions is subsequently measured.

Ion chromatography is a rapid technique with elution time from 1 to 3 minutes per ion, requiring little sample pretreatment other than dilution. It is suited to the analysis of anions (eg, chloride, nitrate, phosphate, sulfate) and cations (eg, calcium, magnesium, potassium, lithium, ammonium, sodium) in biological fluids.

References

1. Willard HH, Merritt LL Jr, Dean TA, Settle FA: Instrumental Methods of Analysis, ed 6. New York, Van Nostrand, 1981, pp 545–550.
2. Snyder LR, Kirkland JJ: Introduction to Modern Liquid Chromatography, ed 2. New York, John Wiley & Sons, 1979, pp 410–438.
3. Pridgar EM, Moses GC, Henderson SR: Purification of lactate dehydrogenase isoenzymes one, two and three from human erythrocytes. Clin Chem 30:1353, 1984.
4. Cirotto C, Allegrucci M, Cartechini MG, et al: Improved column chromatographic isolation of glycosylated hemoglobins. Clin Chem 31:631, 1985.
5. Nicot G, Lachatre G, Gonnet C, Dupuy JL, Valette JP: Rapid assay of N-acetyl-B-D-glucosaminidase isoenzymes in urine by ion-exchange chromatography. Clin Chem 33:1796, 1987.
6. Small H, Stevens TS, Bauman WC: Novel ion exchange chromatographic method using conductimetric detection. Anal Chem 47:1801, 1975.

12
Gel Permeation Chromatography

Separation of molecules according to their size is the basis of gel permeation or exclusion chromatography (Figure 12–1). Gel filtration, molecular sieve chromatography, size-exclusion chromatography, and gel chromatography are additional terms used to describe the technique. In this process, a chromatographic column is packed with a gel having a finite pore size. A sample containing a mixture of constituents with different molecular sizes is then applied. Any molecules larger than the pore size of the gel will be unable to penetrate or permeate the gel; thus, these large-molecule constituents will be eluted first with the eluting buffer. Molecules that penetrate the pore of the gel will be subsequently eluted, with the order of elution depending on the extent of their penetration into the pore of the gel. Thus, small molecules that completely penetrate the pore of the gel will be the last ones to be eluted from the column.

The volume of buffer required to elute molecules unable to penetrate the pore of the gel is referred to as the "void volume" (V_o).[1] The volume of buffer that penetrates the pore of the gel is a reflection of the internal volume of the gel (V_i). The total volume of the gel bed (V_t) is the sum of the volume of buffer occupied by the gel (V_i) and the volume of buffer outside the gel (V_o). Hence;

$$Vt = V_o + V_i \text{ or } V_i = V_t - V_o$$

The volume of buffer required to elute a constituent able to penetrate the pore of the gel (the elution volume, V_e) would depend on the extent of its penetration within the pore or the distribution of the constituent between the internal (V_i), and external volume of the gel (V_o), which is referred to as the distribution, or partition coefficient (K_d). The volume of buffer required to elute a constituent (V_e) can be expressed as:

$$V_e = V_o + K_d V_i$$

Figure 12–1

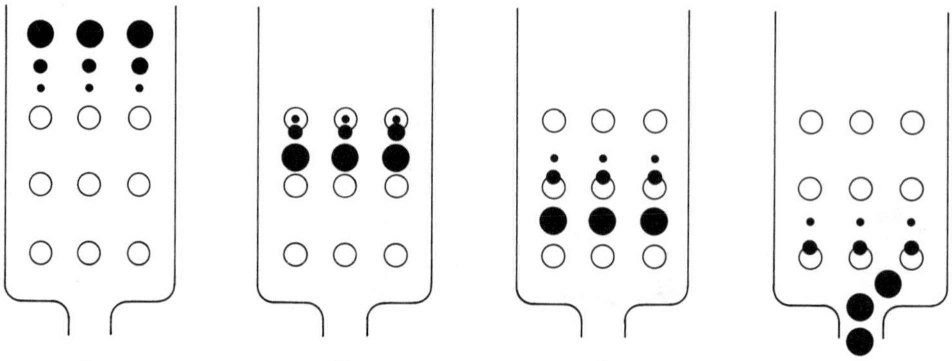

Principle of gel permeation chromatographic process with elution of molecules by size. ○ = Gel beads. Sample constituents: ⬤ = Large molecule too large to enter the pore of the gel, ● = Intermediate molecule that partially enters the pore, • = Small molecule that enters the pore completely. Note: In stage 4, the largest molecule is eluted first, followed by the molecule of intermediate size, and finally by the smallest molecule.

therefore,

$$K_d = \frac{V_e - V_o}{V_i}$$

Molecules too large to penetrate the pore of the gel will be eluted in the void volume (V_o). Molecules that fully permeate the pore of the gel will have a K_d value of 1 and will theoretically be eluted in a volume of buffer twice the void volume. Molecules that partially penetrate the pore of the gel will have K_d values between 0 and 1, depending on the extent of their penetration or distribution within the internal volume of the gel.

In practice, the exact measurement of the internal volume of the gel (V_i) is difficult to determine. Knowledge of the volume occupied by the gel matrix (V_g) is required since

$$V_i = V_t + V_o - V_g$$

and

$$V_t = V_o + V_i + V_g$$

However, in the previous definition of V_t, the volume occupied by the gel matrix (V_g) was regarded as negligible and the V_g term was eliminated in the estimation

Figure 12–2

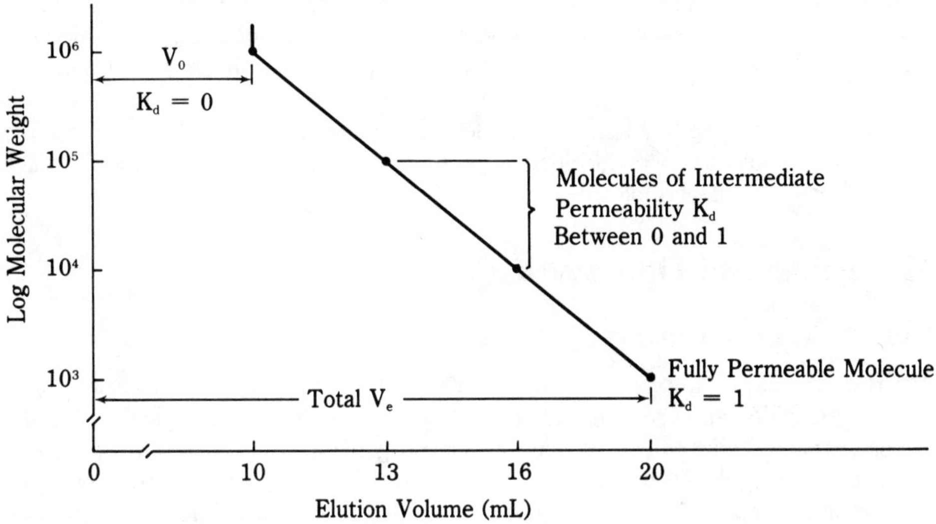

Relationship between molecular size and elution volume. V_0 = volume of buffer (10mL) required to elute the large impermeable molecule with $K_d = 0$, total V_e = total volume of buffer required to elute the smallest fully permeably molecule with $K_d = 1$. The total elution volume is twice V_0 (20 mL). Molecules of intermediate permeability have K_d values between 0 and 1 and are eluted in the order of decreasing molecular size.

of both V_t and V_i. In practice, the distribution coefficient (K_{av}) is calculated without using V_g; therefore,

$$K_{av} = \frac{V_e - V_o}{V_t - V_o}$$

K_{av} values are reciprocally related to molecular size. The smaller the K_{av} value, the larger the molecular size, and vice versa.

Other parameters[2] used in defining the elution behavior of a constituent are the relative elution volume (REV) and the retention constant (R), which is the reciprocal of REV where

$$REV = V_e/V_o \text{ and } R = V_o/V_e$$

Figure 12–2 illustrates the relationship between molecular size and elution volume in gel permeation chromatography.

The effective size of a molecule also depends on its shape, which is a function of its hydrodynamic radius, or Stokes radii. Because precise values of Stokes radii are not readily available for all molecules, molecular weights are used to estimate the size of a molecule. Since the ability of constituents having the same molecular weight to penetrate the pore of a gel will depend on their shape or Stokes radii, molecular weights do not provide an exact estimate of the size of the molecule. Hence, the void volume (V_o) or exclusion volume for constituents of the same molecular weight will vary depending on the shape of the molecule. Thus, globular proteins have a higher exclusion limit than fibrous proteins or linear polysaccharides of comparable molecular weight. The discrepancy can be so large that the elution volume corresponding to a linear molecule of low molecular weight will be the same as for a large molecular weight globular protein whose molecular weight is six times greater than the linear molecule.

Materials and Operation

Gel Packing Materials

A variety of gel packing materials are available. The dextran gels introduced more than two decades ago are still widely used for purification of biological materials. These gels are produced by utilizing the microorganism *Leuconostoc mesenteroides,* which converts sucrose to water-soluble, high-molecular-weight glucose polymer or dextran. The pore size of dextran is regulated by cross-linking the glucose molecules in polymer chains using glyceryl groups introduced during treatment with epichlorohydrin. Cross-linking also renders the dextran gels insoluble in water, although the presence of hydroxyl groups in dextran still permits the gel to take up water and swell. Figure 12–3 illustrates the structure of a portion of a dextran gel. Dextran gels with varying degrees of cross-linking are commercially available and range from highly cross-linked gels, which take up 1 to 2.5 mL of water per gram of dry gel with a molecular weight fractionation range of 700 to 5,000, to the least cross-linked gels, which take up as much as 20 mL of water per gram of dry gel with a molecular weight fractionation range of 5,000 to 800,000. In practice, the lower limit for effective use of these gels is 10% of the stated exclusion limit. These gels have a wide range of particle diameter or bead size. The smaller-bead-size gel with lower particle diameter (10 to 40 μm) and super-fine mesh is ideal when good resolution is desired. Such gels provide sufficient surface area to enable molecules to diffuse more readily into the gel matrix, thereby resulting in good resolution of chromatographic peaks. However, the flow rate of the buffer through small beads will be slow. For preparative chromatographic purposes, the larger-bead-size gels with large particle diameter and coarse (100 to 300 μm) and medium (50 to 150 μm) mesh are adequate. Low cross-linked dextran gels that swell considerably in buffer are soft and cannot withstand high pressures required in high-performance liquid chromatography (HPLC). The highly cross-linked dextran gels, however, are semirigid and can withstand pressures in the range of 150 to 200 psi.

The dextran gels are suitable for use with aqueous buffers. However, the alkylation or acylation of free hydroxyl groups of dextran results in derivatives that can be used with organic solvents.

In addition to dextran gels, polymerized acrylamide beads with varying pore sizes that permit fractionation in the molecular weight range comparable to that of dextran gels are also available. Like the low cross-linked dextran gels, acrylamide

Figure 12–3

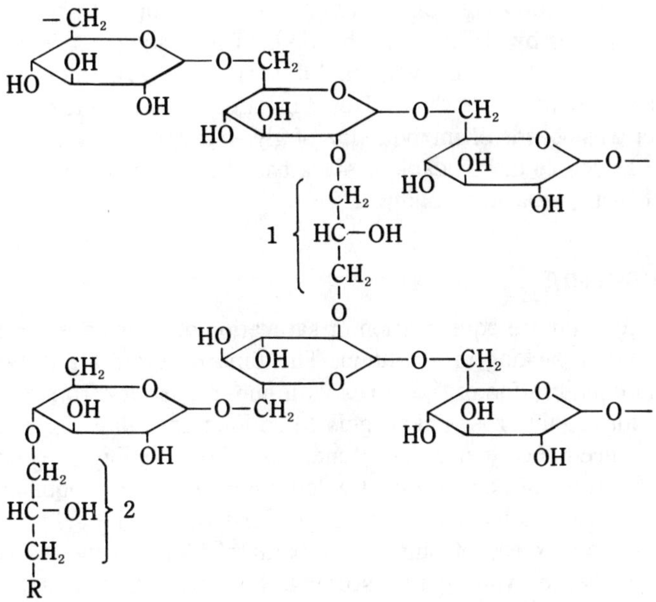

Partial structure of dextran gel. 1 and 2 are cross-linking glyceryl groups linking glucose polymer chains. R = the remaining portion of the dextran molecule.

beads of comparable pore size are not sufficiently rigid to withstand high pressures. Being compressible, the beads slow down the flow rate of buffer through the column, prolonging the time required for the completion of a chromatographic separation.

The introduction of agarose gels has extended the fractionation range for proteins of molecular weight up to 40 million and for polysaccharides up to a molecular weight of 20 million. These gels are provided as a suspension. The agarose gels are stabilized by hydrogen bonding and other forces since they are stable even in presence of hydrogen-bond–breaking agents such as urea or guanidine hydrochloride. The flow rate of buffer with these gels is also high, speeding up chromatographic separations. However, they melt above 40°C, and freezing irreversibly destroys the bead structure.

Thermal stability of agarose gels has been increased by cross-linking with 2,3-dibromopropanol at a highly alkaline pH. The resulting cross-linked agarose also has increased rigidity and yields higher flow rates of buffer than with conventional agarose gels.

Cross-linked styrene-divinylbenzene copolymer particles with particle diameter lower than 10 μm are examples of semirigid gels that have fractionation ranges from small molecules (molecular weight, 50) to large polymers (molecular weight, 10 million) through choice of appropriate pore size, and can be used with most organic solvents with the exception of acetone and alcohol.

The advent of HPLC has resulted in the introduction of rigid gels that can withstand high pressures. Treated porous glass beads or treated silica has been used to speed up chromatographic separations. A mixture of proteins can be separated in 10 minutes by HPLC using bonded-silica packings as compared with the several hours required with conventional dextran gels. Treatment of porous glass beads or silica is required to reduce adsorption effects. Such treatment would involve silanization of glass beads or introduction of glyceryl groups onto the silica matrix.[3]

Rigid gels such as glass beads or silica-based particles used in HPLC are generally available in prepacked columns.

Column Packing

The dextran gels require equilibration or saturation of their pores with buffer prior to de-aeration and packing in a column. The time required for complete swelling of the gel and equilibration of the pores with buffer can vary from 3 hours at room temperature for highly cross-linked gels to as long as 3 days for the least cross-linked gels. These times can be shortened to 1 hour for highly cross-linked gels and 5 hours for the least cross-linked gels by performing the equilibration step in a boiling water bath (100°C). The procedure for swelling the gel consists of sprinkling the gel to an excess of buffer approximately four to five times the water-regain value of the gel while gently stirring. The swollen gel is allowed to settle, the buffer decanted, and the gel resuspended in fresh buffer. The decanting and resuspending of gel in buffer is repeated during the time required for complete equilibration of gel. A magnetic stirrer should not be used for mixing the gel since it will disrupt the gel and produce a large amount of particulate material called "fines." Prior to packing the column, the dextran gel slurry is de-aerated with a vacuum pump. Agarose gels are available as suspension and are ready to be packed in a column after appropriate dilution in buffer.

Columns should be mounted vertically, and it is convenient to use flow adapters to adjust the height of the gel bed (Figure 12–4). The column should be half-filled with buffer. Half of the buffer is allowed to drain to ensure that no air bubble is present in the column or column outlet. The gel slurry is then poured gently with the aid of a glass rod, preferably pouring all the gel required in this step. After the slurry has settled for 10 to 15 minutes, the column outlet is opened and excess buffer in the column is allowed to drain. The excess gel, if any, is removed by aspiration, the top piece or flow adapter is secured, and buffer is passed through the column to equilibrate the gel. Approximately two to three column volumes of buffer are adequate to equilibrate the column. Since dextran gels contain a certain number of carboxyl groups, ion-exchange effects will be evidenced at low ionic strength, which may slow down the elution of positively charged constituents and accelerate the elution of negatively charged species. This effect is overcome by adding sodium or potassium chloride in the range 0.2 to 1.0 mol/L to adjust the ionic strength of buffer between 0.08 and 0.1.

It is important when using low cross-linked soft gels that the operating hydrostatic pressure not exceed that recommended by the manufacturer. The use of a Mariotte flask is recommended to ensure that hydrostatic pressure is constant throughout the experiment.[2]

Figure 12–4

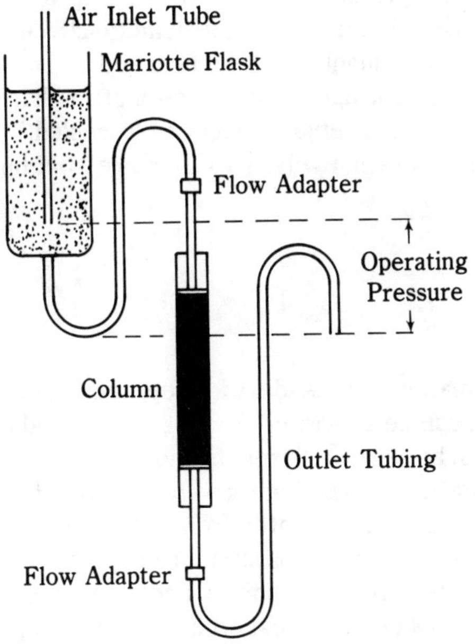

Gel permeation column apparatus. Note: Mariotte flask holds buffer. The flow adapters allow adjustment of column gel bed. Operating pressure is measured as the distance from the bottom of the air inlet tube in the Mariotte flask to the tip of the outlet tubing.

Homogeneity of Gel Packing

The homogeneity of the gel packing is checked by applying a 0.2% solution of blue dextran to the column. If the column has been properly packed, the dye should move down the column in a single band and be eluted in a single symmetrical peak. Because blue dextran has a molecular weight of 2 million, it will be completely excluded by the gel. The amount of buffer required to elute this dye is a measure of the void volume (V_o) of the column.

Application of Sample

Application of the sample can be performed carefully with a precision pipette or applied with the aid of a peristaltic pump that will subsequently also pump buffer through the column. The use of a peristaltic pump will also facilitate maintenance of constant pressure. The sample viscosity should not exceed twice that of the eluting buffer. The flow rate of buffer through the column should be regulated to give optimum resolution since increasing flow rate results in peak broadening, thus decreasing the resolution.

Manipulations

Chromatography can be performed in the ascending direction with buffer being pumped up through the column or in the traditional descending direction. The absorbance of the column eluant can be monitored as the reactions pass through the flow cell of a photometer and displayed on a strip-chart recorder. To improve column efficiency, the column effluent can be "recycled," or passed through the column a second time. Alternatively, two or more columns can be connected in series.

Applications

Gel permeation chromatography is ideal for separating small molecules from large molecules. Free drug can be separated from protein-bound drug. For example, free thyroxine hormone can be separated from bound hormone. Free radiolabeled antigen can be separated from radiolabeled antigen bound to antibody. In labeling experiments, unreacted label can be separated from radiolabeled molecule. Ionized metal such as ionized calcium can be separated from protein-bound calcium and free bilirubin can be separated from bilirubin bound to albumin. Gel permeation chromatography has been applied to clinical studies such as separation of "middle molecule" in uremia,[4] and identification of multiple forms of γ-glutamyl transferase.[5]

The technique is suited to desalting and favored over conventional time-consuming dialysis techniques. The molecular weight of a separated constituent can be determined by utilizing calibrated standards, provided that the standards have the same shape as the unknown constituent. This can be done using a plot of K_{av} vs logarithm of the molecular weight of the calibrator standards.

The introduction of high-pressure gel-permeation chromatography coupled with low-angle laser light scatter detection has facilitated the study of changes in composition of serum proteins in diseased states.[6]

Relative Merits

Gel permeation chromatography is a gentle separation technique that can be used for the separation and purification of labile hormones and proteins. Rapid separation is possible, especially when using rigid gels available for HPLC. Because samples can be injected at definite intervals without risk of overlapping constituent peaks, the technique can be applied to automation including sample injection. The introduction of high-temperature microprocessor-controlled liquid chromatographs has facilitated separation of samples that are soluble only at high temperature, with improved resolution resulting from the lowered viscosity of eluant at high temperature.

The elution of all constituents in a sample in a finite time determined by the void volume is, however, a drawback in that only a limited number of constituents can be separated in a single chromatographic run. In addition, gel permeation chromatography, unlike ion-exchange chromatography, is not suited for separation of isoenzymes of similar molecular size.

References

1. Determan H: Gel Chromatography, New York, Springer-Verlag, 1968, pp 63–65.
2. Cooper TG: The Tools of Biochemistry, New York, John Wiley & Sons, 1977, pp 169–193.
3. Snyder LR, Kirkland JJ: Introduction to Modern Liquid Chromatography, ed 2. New York, John Wiley & Sons, 1979, pp 483–540.
4. Faguer P, Man N, Cueille G, Di Giulio S, Funck-Brentano J: Improved separation and quantification of the "middle molecule" b4–2 in uremia. Clin Chem 29:703, 1983.
5. Wenham PR, Horn DB, Smith AF: Multiple forms of γ-glutamyltransferase: A clinical study. Clin Chem 31:569, 1985.
6. Flapper W, van den Oetelaar PJM, Breed CPM, Steenbergen J, Hoenders HJ: Detection of serum proteins by high-pressure gel-permeation chromatography with low-angle laser light scattering, compared with analytical ultracentrifugation. Clin Chem 32:363, 1986.

13
Affinity Chromatography

Affinity chromatography is a form of adsorption chromatography based on the biological specificity of an immobilized molecule on a solid support (matrix) toward the constituent to be separated. This immobilized molecule, referred to as a "ligand," can be a substrate directed toward a specific enzyme to be purified, an antibody for purification of a specific antigen, or any molecule that can recognize and reversibly bind the desired constituent. The biological specificity of the ligand-constituent interaction confers on this technique tremendous resolving power, since theoretically the desired constituent can be obtained in a highly purified form by merely dissociating the ligand-constituent complex after washing away impurities that have no affinity for the ligand.

Materials and Methods

Matrices

The matrix, or solid support, forms the backbone to which the ligand is attached. Agarose, polyacrylamide, cross-linked dextrans, and porous glass beads are among the many support materials that have been used for affinity chromatography. Of these, agarose has found the widest application. Matrices used in affinity chromatography generally have a gel structure. These gels can be heterogeneous with regions of highly aggregated matrix structure and areas where the matrix structure is absent, creating vacant spaces that are easily accessible to large molecules.[1] Such gels are called "macroreticular gels" and are characterized by high porosity enabling easy access for large molecules into the gel matrix and mechanical stability by virtue of the presence of regions of highly aggregated matrix structure.

In contrast to macroreticular gels, support materials that have a homogeneous structural network are referred to as "microreticular gels." Such gels are less readily permeable to large molecules than macroreticular gels.

Matrix Characteristics

Ideally, the matrix should be chemically inert. There should be no nonspecific adsorption of the desired constituent to the matrix backbone. In practice, the ionic strength of the eluting buffer is maintained above 0.02 to prevent ion-exchange interactions with the matrix. The gel matrix should withstand pressures generated by eluant flow during chromatography. Ideally, this can be achieved by using rigid gels that are also uniform and spherical. The gel matrix should permit optimum flow rate of eluant. Small spherical particles provide good resolution at the expense of a low flow rate. In contrast, large particles permit faster flow rates, with resultant broadening of the applied sample band thus affecting resolution. In practice, the choice of particle size is dictated by the desired degree of resolution and the flow rate. The gel matrix should be stable during the coupling of the ligand to it and withstand an operational range of pH, temperature, and ionic strength encountered during chromatography. The gel matrix should also be resistant to bacterial attack. Finally, one of the most important characteristics of the gel matrix is the ability to couple to it a high concentration of ligand without affecting the mechanical characteristics of the gel (eg, rigidity and porosity).

The greater the ligand concentration the more efficient the binding, and the effective ligand concentration becomes especially important for constituents that exhibit a decreased affinity for the ligand.

Common Matrix Supports

Agarose, controlled pore glass beads, polyacrylamide, and polyacrylamide-agarose gels have been applied in affinity chromatography. Of these, agarose has the widest application. The porosity of agarose gel permits ready accessibility for large molecules to the ligand sites within the gel matrix. Nonspecific adsorption to the gel matrix is also minimized because of the relative lack of charged groups in the agarose matrix. The agarose gels, however, have a narrow range of temperature stability (0°C to 40°C). Freezing irreversibly damages the gel matrix. Their operating pH range is also narrow (between pH 4 to 9), even though they can withstand short-term treatment (2 to 3 hours) to 0.1 M hydrochloric acid or 0.1 M sodium hydroxide. Agarose gels, however, are stable in presence of hydrogen-bond disruptive agents such as 8 M urea or 6 M guanidine hydrochloride, dissociating agents (eg, 3% sodium dodecylsulfate), and salt solutions (1 to 2 M sodium chloride). Prior to use, the agarose gels are stored in the wet state (suspension) in the presence of a suitable bacteriostatic agent, such as 0.02% sodium azide or saturated toluene.

The temperature and operating pH range of agarose gels can be increased by cross-linking agarose with a suitable cross-linking agent, such as 2, 3-dibromopropanol at highly alkaline pH. The resulting cross-linked agarose can withstand autoclaving at 110°C to 120°C and can be used in the pH range 3 to 14 and operating temperature up to 70°C. Unlike ordinary agarose, which is disrupted by organic solvents such as dimethylsulfoxide, the cross-linked agarose can be used with a wide range of organic solvents including dimethylsulfoxide, which broadens its use with a variety of solvents and operating conditions.

The rigidity of controlled pore glass beads permits high flow rates and allows rapid and good separation. The exclusion limits of these beads are governed by the controlled pore size that ranges from 4.5 nm to 250 nm. The beads are stable to autoclaving temperatures and can withstand microbial attack. These glass beads are produced by heating borosilicate glass to 700°C to 800°C to allow separation of the borate and silicate phases. On cooling, the borate phase is dissolved in acid, leaving behind the acid-resistant porous silica glass.

Controlled pore glass beads have a limitation in that the silanol (Si–OH) groups on the glass bead surface have a slight negative charge and absorb positively charged proteins. This disadvantage can be overcome by coating the glass beads with dextran. Although the matrix can be efficiently coupled to a high concentration of ligand, polyacrylamide beads have found limited application because of their low porosity. Polyacrylamide-agarose gels are rigid and permit high flow rates. The size distribution of these beads is narrow (60 to 140 μm), allowing good resolution. The presence of amide groups in polyacrylamide and hydroxyl groups in agarose increases the number of functional groups available for interaction with the ligand.

Choice of Ligand

It is important that the ligand bind readily to its complementary molecule. From a practical standpoint, ligands that have dissociation constants greater than 5 mmol/L in solution are unsuitable.[2] The binding of ligand to the specific constituent to be separated should not be so tight as to require extreme conditions for its dissociation, during which the desired constituent might undergo denaturation. For example, biotin binds to avidin so tightly (dissociation constant, 10^{-15} M) that to dissociate biotin requires elution with 6 M guanidine-HCl at pH 1.5. Under these conditions, most biotin-containing enzymes (carboxylases) are denatured. The ligand should readily bind to the desired constituent and easily dissociate during the recovery step under mild chromatographic elution conditions.

Attachment of Ligand to Matrix

Because agarose is the most widely used matrix for affinity chromatography, only the steps involved in coupling the ligand to this solid support will be discussed.

Prior to coupling of the ligand to agarose, the functional groups in the matrix are activated with cyanogen bromide at pH 11. Addition of cyanogen bromide to agarose beads should be performed under a fume hood because of the toxicity of cyanogen bromide. The suspension should be stirred gently with a glass rod. Magnetic stirrers should be avoided since they will disrupt the agarose gel structure and generate particulate material called "fines." The activation step requires approximately 15 minutes for completion. Because a considerable amount of heat is generated, the mixture should be kept cool by adding ice. Once the activation step is complete, the suspension should be promptly washed with cold sodium bicarbonate or borate buffers (pH 9.5 to 10) followed by the coupling step with the ligand, since

Figure 13–1

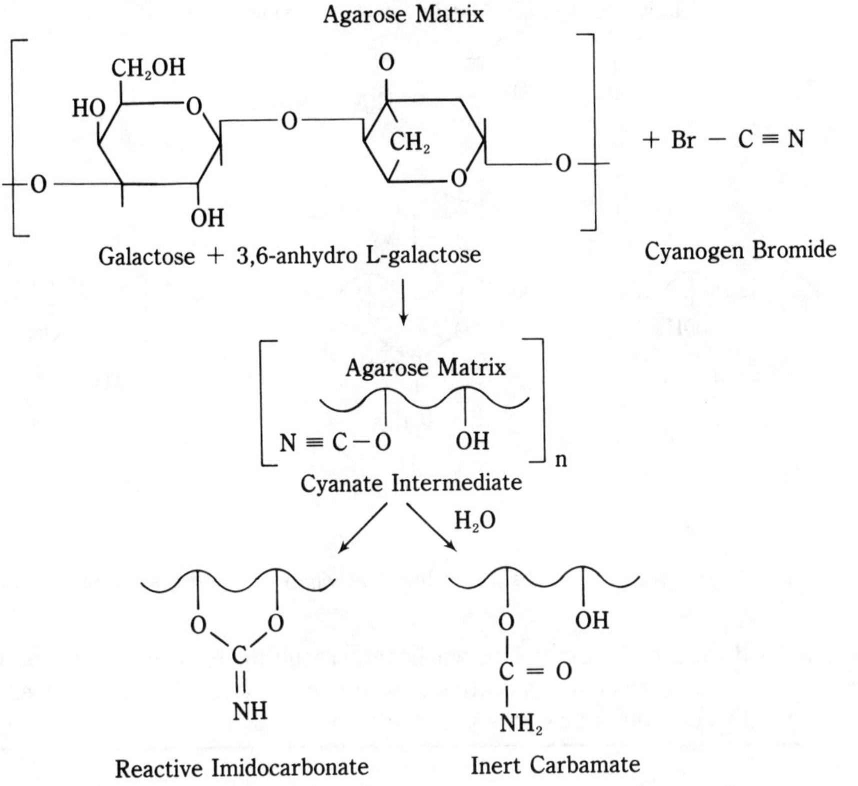

Activation of agarose (a polymer made up of alternating units of D-galactose and 3,6-anhydro L-galactose) by cyanogen bromide to form a labile cyanate intermediate that may either be hydrolyzed to yield inert carbamate groups or react with an adjacent hydroxyl group to form the reactive imidocarbonate species. For simplicity, reactions with hydroxyl group on agarose matrix are shown.

the activated agarose is labile with a half-life of approximately 15 minutes. The washing step should not be performed with buffers containing amino or ammonia groups such as TRIS, glycine, ammonium acetate, or ammonium bicarbonate, because these groups compete with the ligand for attachment to the matrix. The coupling step takes 16 to 20 hours during which time the suspension is gently stirred. The agarose bead is subsequently washed with buffer to remove unbound ligand.

The activation step using cyanogen bromide results in the conversion of hydroxyl groups of agarose to an intermediate cyanate compound that may either be hydrolyzed to form inert carbamate groups, or react with an adjacent hydroxyl group to form both cyclic and acyclic imidocarbonate groups (Figure 13–1). The

Figure 13–2

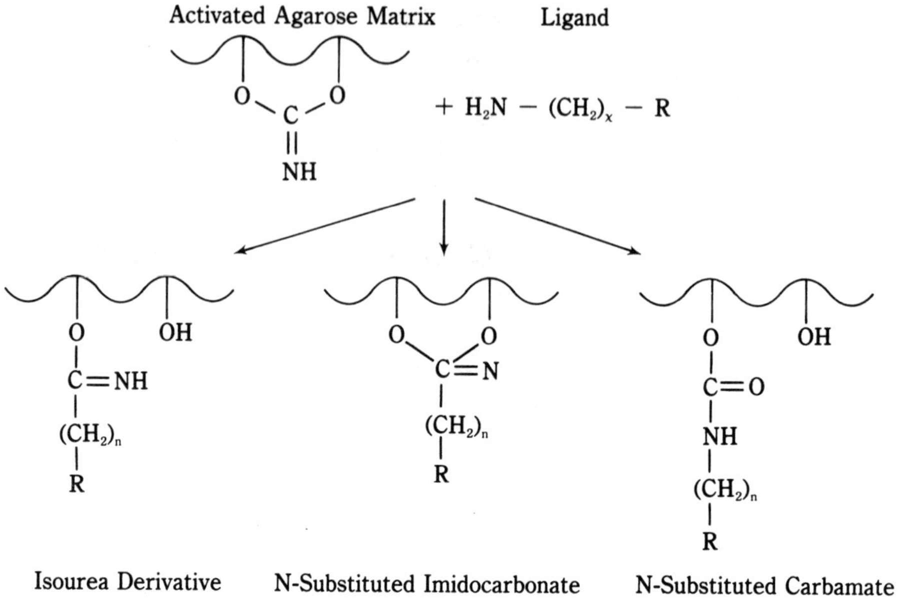

Coupling of ligand to reactive imidocarbonate resulting from cyanogen bromide treatment to yield derivatives indicated. R is the ligand, and $(CH_2)_n$ is a spacer arm. Coupling is through the amino group of the ligand.

gel structure is also rendered thermally stable by the cross-linking produced by formation of acyclic imidocarbonates during cyanogen bromide treatment. The ligand may be coupled through its amino group to the reactive cyclic imidocarbonate group on the activated agarose bead matrix (Figure 13–2). Direct coupling of the ligand to the reactive groups on the agarose matrix is not efficient because access of the desired constituent to the ligand is sterically hindered by the agarose matrix during the chromatography step. This can be circumvented by introducing a "spacer arm" between the agarose matrix and the ligand.[3] The spacer arm is usually introduced by coupling an amine of appropriate chain length to the agarose matrix. The ligand can be coupled to the spacer arm, or the ligand could be the terminal part of the spacer arm itself. The amines used are of the type $NH_2(CH_2)_nR$ where n, the number of methylene groups, can range from 2 to 12. R may represent an amino or carboxyl group to which the ligand is coupled, or it may be the ligand itself. Because the spacer arm should have an appropriate length to avoid steric interference from the matrix during interaction of ligand with the desired constituent molecule, the desired minimum number of methylene groups in the spacer arm should be between 4 and 6. Thus, amines such as hexamethylene diamine ($NH_2(CH_2)_6NH_2$) are applicable in the introduction of spacer arms on the agarose matrix. Treatment with water-soluble carbodiimide is one of the facile organo-

Figure 13–3

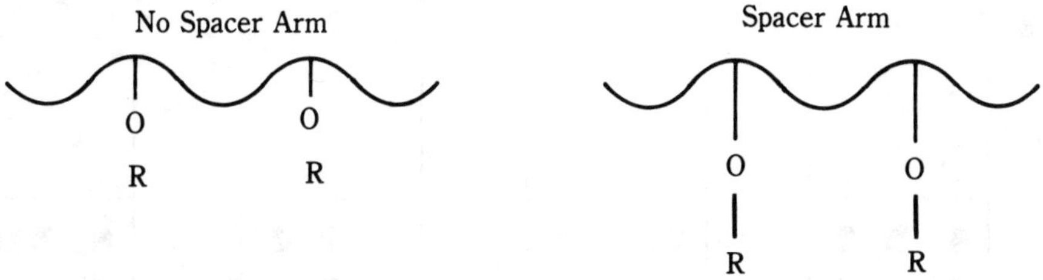

Effectiveness of coupling of ligand (R) with spacer arm (——) inserted between ligand and reactive group (O) on agarose matrix. Without the spacer arm the approach of the ligand to the active group is sterically hindered by the agarose matrix.

synthetic reactions used to couple ligand to the spacer arm. Figure 13–3 illustrates the effectiveness of ligand-constituent interaction with the use of a spacer arm inserted on the agarose matrix.

Procedure

Steps in affinity column chromatography involve equilibration of the affinity gel in buffer (equilibration buffer), adsorption of the constituent to be separated to the affinity ligand (usually done during an incubation step during which buffer flow is interrupted), resumption of flow of buffer to remove unwanted constituents that have no affinity for the ligand, and displacement and elution of the desired constituent.

Typically, a slurry of affinity gel in equilibration buffer is poured into a column containing the same buffer with the column outlet subsequently opened. After the gel bed has formed compactly, the column is connected to the equilibration buffer and the gel bed is washed with approximately ten column volumes of buffer at a controlled flow rate, preferably by using a peristaltic pump, after which the column is ready for sample application. If the constituent to be separated has a high affinity for the ligand, the volume of sample applied to the column is not critical, because the constituent will be tightly bound at the top of the column bed. Constituents having a relatively low affinity to the ligand should be applied in a small volume (50 to 500 μL) to avoid coelution of some of the desired constituent with unwanted constituents that have no affinity for the ligand. The term "progressively perpetuating effectiveness" used by Cuatrecasas[4] best describes the events occurring on sample application in affinity chromatography (Figure 13–4). As sample is applied, a certain fraction of the desired constituent complexes with the ligand. Some of these complexes may dissociate if the binding is weak and add to the desired constituent concentration as the sample migrates through the column bed. This increase in the desired constituent concentration and the accessibility to more ligand binding sites on the gel bed drive the reaction toward complex formation and result

Figure 13–4

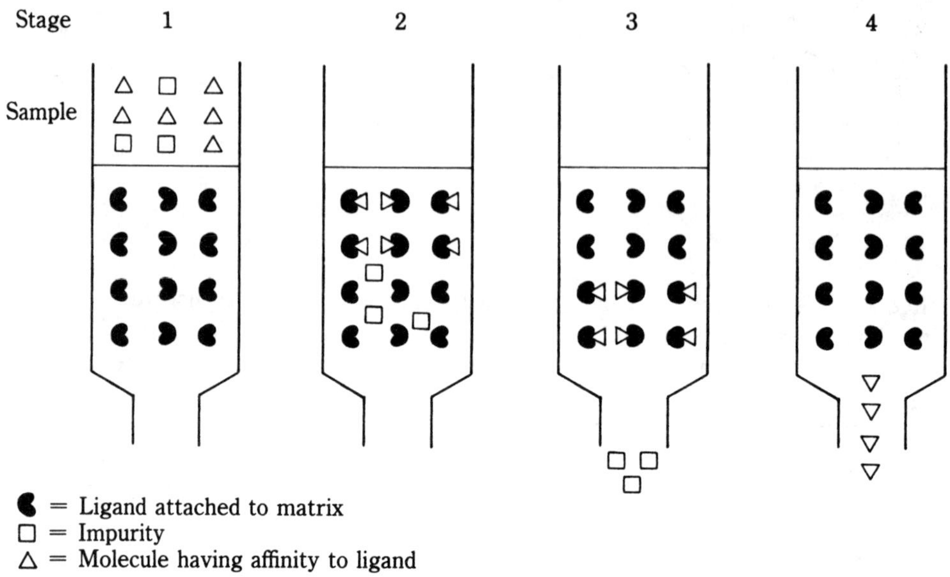

Stages in affinity chromatography. Stage 1: sample application. Stages 2 and 3: binding of molecule to ligand. The impurities do not bind and are washed away by buffer. Stage 4: displacement of molecule from ligand and its subsequent elution.

in effective binding of the desired constituent. To achieve optimum separation in addition to limiting the amount of sample applied, the concentration should also be controlled. For example, when isolating an enzyme protein, the protein concentration of the sample should preferably be kept below 20 mg/mL. High concentrations of protein tend to form aggregates and decrease effectiveness of binding to ligand.

Columns used in affinity chromatography are usually short because the desired constituent is generally bound in the top 1 to 2 cm of the column bed. For all practical purposes, an affinity column can be readily prepared in a Pasteur pipette. As the effectiveness of binding to ligand decreases with increase in temperature, it is preferable to equilibrate the column and perform sample application, incubation, and washing steps (to remove constituents that have no affinity for the ligand) at 4°C.

Subsequent to incubation and washing steps to remove unbound constituents, elution of the desired constituent can be accomplished in a variety of ways. The constituent may be eluted by increasing the temperature, the pH, or the ionic strength of the buffer. Alternatively, a competitor or inhibitor can be used to displace the desired constituent from the ligand. Gradient elution techniques have been used to effect increases in pH, temperature, ionic strength, or co-enzyme concentration. Thus, the isoenzymes of lactate dehydrogenase were eluted with a gradient of NADH.[5]

The use of low pH and hydrogen-bond breaking agents such as urea or guanidine-hydrochloride may be required in dissociating proteins that bind extremely tightly to the ligand. In such instances, it is important to restore the optimum pH required for the stability of the eluted protein and also to remove or neutralize the hydrogen bond-breaking additives. An alternative to affinity column chromatography is the batch procedure, which lends itself readily to constituents that have a strong binding to ligand. In such a procedure, the sample is added to a slurry of adsorbent gel. After an adequate incubation period to effect optimum binding to the ligand, the adsorbent gel is centrifuged, the supernatant discarded, and the constituent bound to the gel sediment displaced and recovered by adding appropriate eluant buffer. Batch procedures are also useful to determine the incubation times required for optimum binding in the affinity chromatography step.

Other Affinity Techniques

In classic affinity chromatography, the ligand-specific constituent complex is reversible. In a specialized form of affinity chromatography called "covalent chromatography," the desired constituent is trapped in a covalent linkage with the ligand. The desired constituent is subsequently released by appropriate chemical treatment such as the use of a reducing agent.[6] Covalent chromatography has found application for purification of proteins containing thiol groups such as creatine kinase, phosphofructokinase, and urease.

In metal-chelate affinity chromatography (MCAC), a metal chelate such as zinc or copper chelate is linked to derivatized agarose. Proteins bind to metal by means of exposed histidine, cysteine, or tryptophan residues. Since the binding is usually at an optimum pH, variation of pH recovers the adsorbed protein. Alternatively, a metal chelant (eg, EDTA) can be used to displace the adsorbed protein from the metal site.[7] MCAC has found application in the purification of α_2-macroglobulin, plasminogen activators, and interferon.

Cell surface carbohydrates can be visualized by affinity histochemistry techniques. This can be accomplished by incubating cells with a phytohemaglutinin such as concanavalin A, which binds to mannose residues on the cell. Because lectins such as concanavalin A have two sugar-binding sites, they can also bind a histochemical marker such as ferritin containing covalently bound mannose groups or glycosylated horseradish peroxidase.[8] The ferritin complex can be examined by electron microscopy. The peroxidase complexed cell surface carbohydrates can be seen under light microscopy.

Applications

Affinity chromatography has been used for the separation of a variety of molecules of clinical interest ranging from enzymes, isoenzymes, specific proteins, glycoproteins, coagulation factors, hormones, receptor proteins, and cells to interferon. A recent application is in the routine measurement of glycosylated hemoglobin. The basis of this measurement is the immobilization of the stable glycosylated hemoglobin (ketoamine) fraction to boronic acid derivatized agarose and subsequent elution with a competing sugar (sorbitol) in the eluant buffer.

The affinity chromatographic procedure, unlike ion-exchange procedures, can be performed without pretreatment of the erythrocytes to remove the labile glycosylated hemoglobin fraction (Schiff base) prior to preparing the hemolysate and can withstand moderate variations in temperature and pH.[9] The properties of protein A isolated from the cell wall of *Staphylococcus aureus* have been exploited in affinity chromatography for the separation of immunoglobin (IgG) subclasses.[10] (IgG subclasses 1, 2, and 4 and their F_c region fragments in serum bind to protein A; IgG_3 does not.) The demonstration of enzyme-IgG complexes in serum such as creatine kinase-IgG complex, and alkaline phosphatase-IgG complex can be performed by affinity chromatography on agarose to which protein A is coupled.[10] Affinity chromatography of carcinoembryonic antigen (CEA) and α-fetoprotein (AFP) on concanavilin-A derivatized agarose has demonstrated the existence of two forms of CEA and AFP. The differential quantitation of the fraction of AFP that contains the sugar mannose is of value in the diagnosis of neural tube defects.[11] Affinity chromatography of serum alpha-fetoprotein on lentil lectin has been used in the differential diagnosis of benign and malignant liver disease, since patients in the latter category have a greater percentage of lentil lectin-reactive serum alpha-fetoprotein.[12] The mini-column affinity chromatography procedure for the separation of alpha-fetoprotein, using as little as 100 μL of serum, lends itself to the routine analysis of serum from cancer patients.[13] Purification of isoenzymes of lactate dehydrogenase,[14] cancer-associated galactosyltransferase isoenzyme,[15] and placental-type alkaline phosphatase in testicular cancer[16] have been performed utilizing affinity chromatography. The current interest in glycated albumin to assess short-term glucose changes in pregnant diabetics is an area where affinity chromatography has been exploited in the clinical laboratory.[17]

The voluminous literature on the subject demonstrates the usefulness of affinity chromatography to the clinical scientist.

References

1. Lowe CR: An introduction to affinity chromatography. In Work TS, Work E (eds): Laboratory Techniques in Biochemistry and Molecular Biology, New York, North-Holland, 1979, pp 293–358.
2. Cooper TG: The Tools of Biochemistry. New York, John Wiley & Sons, 1977, pp 234–55.
3. Cuatrecasas P: Protein purification by affinity chromatography. Derivatizations of agarose and polyacrylamide beads. J Biol Chem 245:3059, 1970.
4. Cuatrecasas P: Affinity chromatography of macromolecules. In Meister A (ed): Advances in Enzymology, vol 36. New York, John Wiley & Sons, 1972, p 29.
5. Brodelius P, Mosbach K: Separation of the isoenzymes of lactate dehydrogenase by affinity chromatography using an immobilized AMP-analogue. FEBS Lett 35:223, 1973.
6. Brocklehurst K, Carlsson J, Kierstan MPJ, Crook EM: Covalent chromatography. Preparation of fully active papain from dried papaya latex. Biochem J 133:573, 1973.

7. Porath J, Carlsson J, Ohlsson I, Belfrage G: Metal chelate affinity chromatography, a new approach to protein fractionation. Nature 258:598, 1975.

8. Kieda C, Delmotte F, Mosigny M: Preparation and properties of glycosylated cytochemical markers. FEBS Lett 76:257, 1977.

9. Klenk DC, Hermanson GT, Krohn RI, et al. Determination of glycosylated hemoglobin by affinity chromatography: Comparison with colorimetric and ion-exchange methods, and effects of common interferences. Clin Chem 28:2088, 1982.

10. Baur K, Bayer PM, Deutsch E, Gabl F: Binding of enzyme-IgG complexes in protein-A sepharose CL-4B. Clin Chem 26:297, 1980.

11. Buamah PK, Taylor P, Ward AM: Concanavalin A binding of α-fetoprotein in amniotic fluid as an aid in the diagnosis of neural tube defects. Clin Chem 27:1658, 1981.

12. Buamah PK, Harris R, James OFW, Skiller AW: Lentil-lectin-reactive alpha-fetoprotein in the differential diagnosis of benign and malignant liver disease. Clin Chem 32:2083, 1986.

13. Chan DW, Miao Y: Affinity chromatography of alpha-fetoprotein variants: Development of a mini-column procedure, and application to cancer patients. Clin Chem 32:2143, 1986.

14. Pridgar EM, Moses GC, Henderson AR: Purification of lactate dehydrogenase isoenzymes one, two, and three from human erythrocytes. Clin Chem 30:1353, 1984.

15. Ram BP, Munjal DD: Isolation and characterization of cancer-associated galactosyltransferase isoenzyme. Clin Chem 30:1656, 1984.

16. Shameem GMM, Quadri F: Isolation and purification of placental-type alkaline phosphatase from a seminoma. Clin Chem 33:248, 1987.

17. Rendell M, Kao G, Mecherikunnel P, et al: Aminophenylboronic acid affinity chromatography and thiobarbituric acid colorimetry compared for measuring glycated albumin. Clin Chem 31:229, 1985.

14
Gas Chromatography

Partition of volatile constituents in a sample between a liquid coated on a solid support (stationary) phase and a mobile gas phase is the basis of gas-liquid chromatography (GLC). When this partition occurs between a solid stationary phase and a mobile gas phase, the technique is referred to as "gas-solid chromatography" (GSC). Because both GLC and GSC are based on the same principle, the term "gas chromatography" (GC) will be used to describe the partition of sample constituents between a mobile gas phase and a stationary phase. For a constituent to be separated by gas chromatography, it must be volatile at the temperature prevailing in the chromatographic column. If the sample constituent is not volatile in its present form, it must be capable of being transformed to a volatile derivative.

Design and Components

Figure 14–1, a schematic representation of a gas chromatograph, shows an injection port for injection of a sample into a column inlet. The injection port is heated to a temperature at which the volatilization of the injected sample occurs, at which point it is referred to as the "flash heater." The injected sample is swept through the column packing by an inert gas. Gas tanks and pressure regulators are required for the carrier gas. The column filled with column packing is encased in a thermostatted oven to maintain the column at the desired temperature.

The column packing consists of a liquid stationary phase coated on a solid support. Selected solid supports can also serve as a stationary phase. Detectors heated to a desired temperature sense the emergence of constituents of a sample as they arrive at the column outlet. The electrometer monitors the appropriate signal from the detector. The emergence of sample constituent peaks at an appro-

Figure 14–1

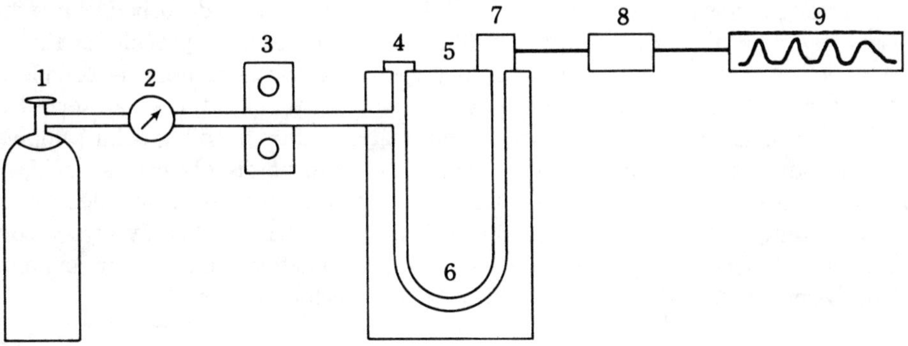

The components of a gas chromatograph: (1) carrier gas cylinder, (2) gas pressure regulator, (3) gas flow regulators, (4) injection port, (5) thermostated oven, (6) column, (7) detector, (8) electrometer, (9) recorder.

priate time during chromatography (retention time) is recorded on a strip-chart recorder. Depending on the type of detector, additional gas tanks and pressure regulators may be required.

Injection Port

The sample is injected through a heated self-sealing rubber septum. The heated septum of the injection port, or flash heater, is maintained at a temperature generally 30 °C above the temperature of the column. The sample is volatilized upon injection through the flash heater with the volatilized sample constituents swept through the column packing by a stream of carrier gas. Some injection ports allow direct injection of sample into the column. "On-column" injection procedures minimize loss of sample by interaction or adsorption to dead space between the injection port and the column. For constituents that are thermally labile and unable to withstand flash evaporation, injection of samples directly onto the column is especially important prior to their being swept into the column by a preheated carrier gas stream.

Sample size and concentration are critical for good peak resolution. A large sample size or a concentrated sample will result in overloading the column; this leads to asymmetrical peaks on the gas chromatogram. For liquid samples, the preferred sample size ranges from 1 μL to 10μL. In a 1-μL sample, the concentration of each sample constituent should be approximately 1% of the sample injected to ensure good peak resolution.

Gas samples are generally injected into the gas chromatograph with a gas-tight syringe. However, gas samples can be injected using a calibrated sample loop and a variety of leak-proof valves.[1,2] Automated sample loop and valve arrangements enable automated injection of liquid samples. When a small sample size is required, as with capillary columns, the sample introduced by a syringe is split, allowing only

0.01 of a 1-μL sample to enter the column. Devices are available for trapping volatile constituents of a biological or environmental sample that can be readily transported to the site of analysis. The trapped constituents are desorbed by passing carrier gas at a high temperature (300°C) and collected in a precolumn that is maintained at a low temperature with dry ice. Once the desorption is complete, the contents of the precolumn are brought to ambient temperature, transferred to the column of the gas chromatograph, and analyzed. For handling solid samples, pyrolytic devices are preferred. The solid sample can be dissolved in a liquid, or melted and placed on a wire, or deposited in solid form on a probe and electrically heated to temperatures as high as 800°C within nanoseconds. The pyrolyzed constituents are subsequently swept through a gas chromatographic column and analyzed. Recently, pulsed laser beams have been used for pyrolysis.[3]

Carrier Gas Cylinders

Carrier gas cylinders equipped with flow meters are used to control gas pressure, maintaining the carrier gas flow at the required rate through the column. Helium or nitrogen is the carrier gas usually used. An oxygen-water scrubber is placed on the carrier gas line to remove moisture, and a new one should be used with each new tank. Depending on the type of detectors used, additional gas cylinders and pressure regulators may be needed. Flame ionization detectors use hydrogen gas, and electron capture detectors use argon or methane gas.

Column

Columns packed with a stationary phase coated on a solid support can be made of metal (eg, copper or aluminum). As some constituents such as steroids may react with the oxide films of these metals, columns of stainless steel are preferred. Columns made of glass are ideally more inert than metallic columns. Metallic or glass columns, in the form of coils, are generally 1 m to 3 m long, with an internal diameter of 2 mm to 9.5 mm. Glass columns can be coiled or "U" shaped. Open tubular or capillary columns are generally 50 m to 150 m long. Capillary columns having an inner narrow wall of approximately 0.25 mm and coated with a liquid stationary phase of approximately 1 μm in thickness are referred to as wall-coated open tubular columns (WCOT).[4]

The WCOT columns, however, have limited sample capacity. An improved column, the support-coated open tubular column (SCOT) has been devised. In a SCOT column, a porous support layer coated with a liquid phase is either chemically bonded to or deposited on the inner wall of the capillary tubing. As the liquid phase is well dispersed in the porous support layer of the SCOT column, the resulting film thickness is lower than in the WCOT column. In addition, the SCOT column has a greater amount of liquid phase per unit length of the column when compared with the WCOT column. SCOT capillary columns have increased sample capacity. Because of the increase in the number of theoretical plates, the desirable column length can be considerably shorter when compared with WCOT columns.

Currently available are high-efficiency capillary columns with a coating of cross-linked silicone that can be up to 0.5 μm thick, greatly increasing sample capacity and providing good peak resolution and reducing analysis time. Increasing the internal diameter of capillary columns from 0.25 mm to 0.5 mm not only allows the coating of a stationary phase as thick as 2 μm, but permits use of a large sample size by direct injection. Such wide-diameter columns are useful for identification of sample constituents by a coupled mass spectrophotometer, although the efficiency of such stationary phases is comparably low and the analysis time is long.

Capillary columns are usually made of fused silica. Because of the length of capillary columns, the column efficiency (as judged by the number of plates) can be as high as 600,000 plates as compared with 10,000 to 20,000 plates achieved with stationary phases in conventional packed columns. Capillary columns have the additional advantage of providing the carrier gas an unrestricted flow through the column, resulting in a relatively small pressure drop. Capillary columns are usually coiled in a spiral.

Packing conventional or capillary columns is an involved procedure requiring practice. Many people prefer to purchase prepacked columns. Prior to packing a conventional column, the liquid stationary phase is coated on a solid support by using the rotating evaporator procedure. In this procedure, the appropriate amount of liquid phase is dissolved in a solvent in a round-bottom flask to which the required quantity of solid support is added. The flask is rotated until all of the solvent is removed under reduced pressure, and the firmly coated stationary phase on the solid support is then ready for packing. The next step is to place a funnel at one end of the column and to plug the other end with glass wool. Tubing that allows suction to be applied through a water aspirator can be inserted into the end with the glass wool. The column packing material is poured through the funnel into the column.

After the column is completely filled, it is vibrated either by a hand vibrator or by tapping with a stick until the packing material is well packed. The funnel is removed, and the open end plugged with glass wool. Coating of capillary columns is done either by forcing a liquid phase suspension through the column with the aid of an inert gas, or by completely filling the column tubing with liquid phase suspension and allowing the solvent to evaporate from one open end of the column. Prior to use, columns should be conditioned in the temperature-controlled column oven at a temperature at which the stationary-phase liquid is still nonvolatile.

Depending on the stationary phase selected, the conditioning time and temperature will vary. The columns should be conditioned with a small flow of carrier gas. During the conditioning step, the column should not be connected to the detector end in order to avoid contaminating the detector.

Stationary Phases

Stationary liquid phases used in gas chromatography should be nonvolatile at the temperature selected for analysis. The stationary phase liquid should be stable, not interact chemically with the sample constituents, and permit selective partitioning of sample constituents. The operating column temperature is usually 10°C to 15°C below the upper temperature limit for the stationary phase to prevent volatilization of stationary phase (bleeding) and contamination of the detector. As separation occurs only in the liquid phase, the choice of a stationary phase should depend on the relative solubility of sample constituents in the liquid stationary phase. The mobile carrier-gas phase is inert and merely serves to sweep the sample constituents to the detector depending on their affinity for the stationary phase. Nonpolar samples can be separated on stationary phases that are methyl-substituted silicone polymers referred to as SE-30, OV-1, or OV-101. These methyl-substituted silicone polymers have an operating temperature limit at 325°C. These polymers separate hydrocarbons based on their carbon number, alcohols by the differences in boiling points, and fatty acids according to their carbon number rather than the degree of unsaturation. These polymers are referred to as "nonselective liquid phases."

Stationary phases that separate on the basis of polarity are called "selective stationary phases." Examples of selective phases are polyethylene glycol (carbowax) for separation of hydrogen-bonded species such as alcohols, neopentyl glycol succinate (NPGS) for fatty acids, fluoroalkyl silicone (QF-1) for separation of fluorinated derivatives of amino acids, cyanoethylmethyl silicone (XE-60) for separation of drugs with nitrogen function, and methyl phenyl silicone (OV-17), a widely used stationary phase for the separation of polar compounds. Polar stationary phases are stable up to 225°C with the exception of methyl phenyl silicone, which is stable to 300°C. Cross-linked polar and nonpolar silicones have been used as stationary phases in high-efficiency capillary columns.

Column Packing Supports

Column packing supports provide a large surface area for coating of the liquid stationary phase. Widely used column packing supports include the diatomaceous earths, which have small, uniformly sized particles composed mainly of hydrated silica with metal oxide impurities in trace amounts. As these silica supports are reactive, they should be treated prior to coating with liquid stationary phase. Acid washing removes mineral impurities that would otherwise adsorb the sample. The active silanol groups (Si–OH) that can adsorb sample constituents are neutralized by silanization with dimethyldichlorosilane (DMCS), converting the silanol groups to silyl ether groups. The excess dimethyldichlorosilane reagent remaining is subsequently neutralized with methanol. Neutralization of silanol groups on the support eliminates peak trailing that would result because of hydrogen bonding of the silanol groups with polar sample constituents. Treatment of diatomaceous earth with either DMCS or hexamethyl disilizane (HMDS) converts both silanol and siloxane (Si–O–Si) groups to silyl groups, preventing hydrogen bonding with sample constituents.

Diatomaceous earth supports prepared from firebrick are designated by various brand names (eg, Chromosorb W, P, A, T, and G [Johns-Manville Corporation]). The supports are sized by their ability to pass through standard mesh screens designated by the American Society of Testing Materials. Recommended support particle size is 100/120 mesh (indicating that particles will pass through 100 mesh but not through 120 mesh) or 149μm to 125μm for 2-mm internal diameter columns, and 80/100 mesh (177μm to 149μm) for 4-mm internal diameter columns. The concentration of a liquid-stationary phase coated on a solid support usually ranges from 2% to 10% weight/weight depending on the stationary phase. Chromosorb G, W, and T are used to coat stationary phases for separation of polar compounds, and chromosorb P is used for coating stationary phases for separation of hydrocarbons. Chromosorb A, on which a liquid phase as high as 25% can be coated, is used mainly for preparative gas chromatography. The adsorptive properties of the untreated support can be exploited to separate gases and low–molecular-weight hydrocarbons by gas-solid chromatography (GSC). Silica gel, charcoal, and molecular sieves such as sodium aluminum silicate (synthetic zeolites) are some of the supports that have been used in GSC.

Detectors

Detectors having wide applications in gas chromatography include the thermal conductivity detector, flame ionization detector, nitrogen-phosphorus detector, and electron capture detector. The temperature of the detector compartment is approximately 10% over the column temperature (°C) to prevent condensation of the sample. The thermal conductivity detector is based on the principle that a heated element will lose heat during the passage of a gas over it. The amount of heat lost will depend on the thermal conductivity of the gas. Carrier gases such as helium have a relatively high thermal conductivity. When a carrier gas stream mixed with sample constituents flows over a heated filament, its thermal conductivity is reduced, the filament retains more heat, and its electrical resistance increases because of the increase in temperature of the filament. A thermal conductivity detector has two or four heated filaments of either tungsten wire or a tungsten-alloy wire containing 3% rhenium. Carrier gas alone flows over one of the two filaments (the reference filament), or over the two filaments in a four-filament arrangement that correspond to the reference filament. Carrier gas flowing through the column and carrying sample constituents flows over the one remaining filament in a two-filament arrangement (the test filament) or over the two remaining test filaments in a four-filament arrangement. When carrier gas is flowing over the reference filament, and carrier gas without sample constituents is emerging from the column and flowing over the test filament wires, both wires will have the same temperature and resistance, and the circuit will be balanced. When carrier gas mixed with sample constituents from the column flows over the test filaments, the resistance of the reference and test filaments will change because of the difference in temperature of the reference and test filaments, and the circuit will remain unbalanced. The difference in resistance between the reference and test filaments is measured and is related to the concentration of the sample constituent emerging with the carrier gas from the column. The sample constituents emerging from the column onto the thermal conductivity heated filament can be collected for further analysis by other techniques. Thus, the sample constituents are essentially nondestructive. The thermal conductivity detector has moderate sensitivity and a detection limit of approximately 5μg/mL to 10μg/mL of carrier gas. The flame ionization detector uses the temperature of the flame to convert sample constituents emerging from the column to ions and free electrons (Figure 14–2A). The flame gases are hydrogen and air, or oxygen. Above the flame jet is a platinum loop that serves as a collector electrode. A voltage of approximately 400 V applied between the flame jet and collector electrode lowers the resistance between these two electrodes, ensuring current flow. As sample ions are collected at the collector electrode, the resulting current is amplified by the electrometer, and the signal is displayed on the strip-chart

Figure 14–2

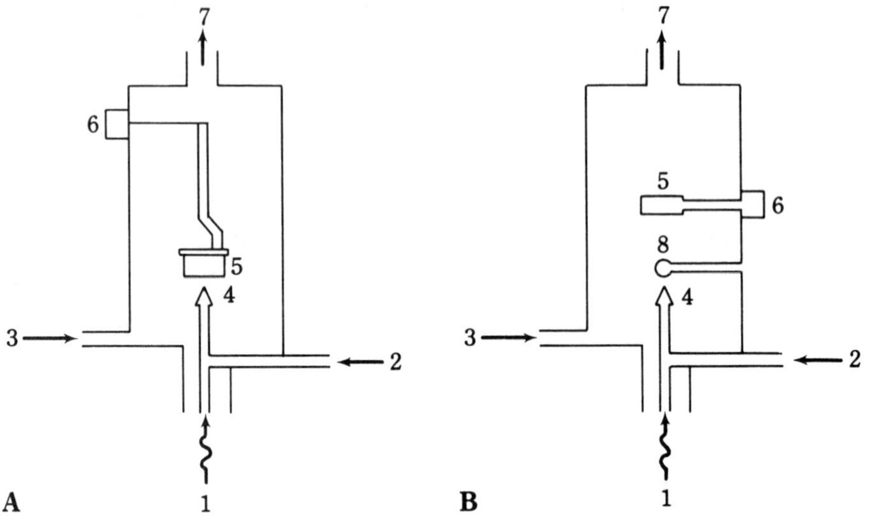

A: Flame ionization detector: (1) carrier gas from column, (2) hydrogen gas inlet, (3) air gas inlet, (4) flame jet, (5) collector electrode, (6) electrometer receiving impulse from collector electrode, (7) vent. B: Nitrogen-phosphorus detector. All components are the same as in Figure 14–2A with the addition of a rubidium silicate bead (8) between the flame jet and collector electrode.

recorder in the form of a peak. The flame ionization detector does not respond to water or to gases such as nitrogen, oxygen, carbon monoxide, carbon dioxide, sulfur dioxide, ammonia, hydrogen sulfide, nitrous oxide, silicon tetrafluoride, silicontetra chloride, and carbon disulfide. This property can be exploited in measuring organic compounds containing gaseous impurities.

The flame ionization detector is not particularly sensitive to oxygenated carbons such as carbonyl or carboxylic acid groups, thio groups, and ethers. The response of this detector is proportional to the number of carbon atoms. Increasing substitution by amines, halogens, hydroxyl groups, or any electron-capturing species reduces the flame ionization detector response. The flame ionization detector has increased sensitivity compared with the thermal conductivity detector and has detection limits of approximately 1 ng/mL to 5 ng/mL of carrier gas.

A modification of the flame ionization detector is the nitrogen-phosphorus detector or alkali emission detector. Using low hydrogen and air flow rates, 10 mL/min and 100 mL/min, respectively, keeps the temperature of the flame low enough to suppress the ionization of compounds other than those containing nitrogen or phosphorus. A glass bead containing a nonvolatile alkali salt such as rubidium silicate is placed between the flame jet and the collector electrode (Figure 14–2B). The bead is electrically heated between 600°C and 800°C. Sample constituents containing nitrogen or phosphorus are pyrolyzed at the bead surface, and the resulting radicals ($\cdot C = N$ or $\cdot P = 0$) capture electrons from the thermally excited rubidium atom to form ions that migrate to the collector electrode to produce a signal response. The positive rubidium ion produced in the process is captured by the bead because of the negative potential that has been applied to it; it is there ready again to undergo the cycle of reaction with sample radicals generated at the bead surface. The nitrogen-phosphorus detector is approximately 50 times more sensitive to nitrogenous compounds and approximately 500 times more sensitive to phosphorus-containing compounds when compared with the flame ionization detector. By appropriate variation of flow rates of hydrogen and air, the nitrogen-phosphorus detector can be made selective to either nitrogen or phosphorus compounds.

The electron capture detector is sensitive to species deficient in electrons. A radioactive source such as tritium or nickel 63 is used. ^{63}Ni foil is preferred because it can be heated to 400°C. In contrast, a platinum foil saturated with tritium cannot be heated above 225°C. A carrier gas-mixture such as helium-methane, argon-methane, or nitrogen-methane is passed over the radioactive source. The beta particles ionize the carrier gas, and the resulting electrons migrate to the anode under an applied voltage. The electrons produce a steady current that is amplified by the electrometer. These electrons are captured by sample constituents deficient in electrons, causing a reduction of current proportional to the concentration of the electron-deficient species in the sample. The electron capture detector is particularly sensitive to halogenated compounds, conjugated carbonyls, anhydrides, peroxides, nitriles, nitrates, organometallic and sulfur-containing compounds; however, it is insensitive to alcohols, amines, ketones, and hydrocarbons. The electron capture detector is a very sensitive detector, achieving a picogram-level detection limit. As this detector also responds to oxygen and to water vapor, it is important that the carrier gases selected are free of oxygen and the moisture that produces unstable baselines on the recorder.

Additional Components

The electrometer's function is to amplify the signal from the detector. The signal can be displayed on a strip-chart recorder. The advent of the microprocessor has simplified data reduction in gas chromatography. The microprocessor-controlled electronic integrators amplify the detector signal and provide a digital printout of peak areas and retention times. Baseline correction and determining areas of fused peaks are among the functions that can be performed by the electronic integrator, which has a wide linear range. These devices also assign the area percentage of each peak relative to the sum of areas of all peaks in the chromatogram.[5]

Because the identification of each sample constituent peak in gas chromatography is based on its retention time, it is important that retention times be reproducible from run to run. Reproducibility of retention times depends on a strict control of column and detector temperature and carrier gas flow rates. For quantitative analysis, precise sample injection volumes should be maintained from run to run. In practice, it is difficult to control precisely all these variables, and an internal standard is generally included with a sample. By measuring the peak area ratio of the sample constituent peak to the internal standard peak, changes in sample injection volume and variables such as carrier gas flow rates and column and detector

Figure 14–3

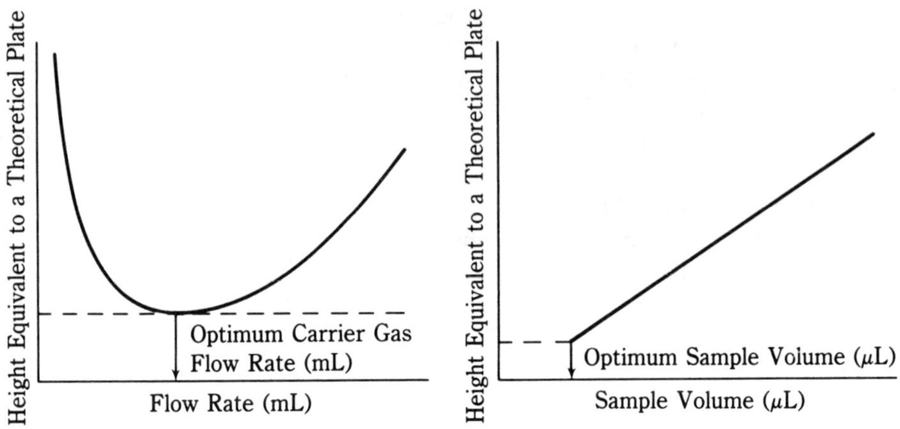

Calibration plots for the optimization of gas chromatographic conditions.

temperature will affect the sample and internal standard to the same extent and will be compensated. The internal standard chosen should not interact with any of the sample constituent peaks, must be well resolved from sample constituent peaks, and must elute close to the sample constituent peak of interest. The internal standard should also be structurally similar to the sample constituent peak of interest, and its response should be linear in the desired range for sample constituent concentration.

Conditions

Conditions such as the temperature of the flash heater and detector and the sample volume should be stated for each chromatogram. Chromatography can either be performed at a fixed-column temperature (isothermal) or be programmed to increase at a fixed rate to the desired final temperature (temperature programming). Whereas peak sharpness would depend on column efficiency (the number of plates), in general, peak broadening of constituents that emerge toward the end of a chromatographic run occurs under isothermal conditions. Peak broadening can be overcome by using temperature programming that also shortens the time ordinarily required for chromatographic analysis under isothermal conditions. Optimum conditions for gas chromatography (Figure 14–3) are established by a plot of height equivalent to a theoretical plate vs carrier gas flow rate (A) or sample volume (B). The optimum carrier gas flow rate or sample volume is that which can be attained at the lowest height equivalent to a theoretical plate (Figure 14–3).

Samples

Samples that are nonvolatile under the temperature conditions used in gas chromatography are rendered volatile by being broken down into volatile fragments by pyrolysis or derivatization. Depending upon the chemical nature of the sample, a variety of derivatization techniques are available. Fatty acids can be converted into methyl esters by methylating with methanol in the presence of boron trifluoride as a catalyst. The methyl esters are extracted with a suitable solvent (eg, petroleum ether) and injected into the gas chromatograph. Silylating agents can be used to convert an active hydrogen in a hydroxyl, carboxyl, or amino group to a trimethylsilyl ether [$-O-Si(CH_3)_3$] group. Carbohydrates and steroids can be derivatized as trimethylsilyl ether derivatives. Amino acids can be converted to *N*-trifluoroacetyl-alkyl esters that are easily analyzed with an electron capture detector.

Catecholamines can be converted to their perfluorederivatives. Barbiturates can be converted to their dimethylderivatives with either tetramethylammonium hydroxide or trimethylanilinium hydroxide directly on the column (on-column derivatization) by injecting either of the hydroxides simultaneously with the sample.

Applications

In addition to the analysis of fatty acids, steroids, carbohydrates, amino acids, and catecholamines, gas chromatography has wide applications in the field of toxicology. The technique of head space analysis has been particularly useful to assay alcohol in blood samples.[6,7] In this technique, an aliquot of blood is mixed with salt (sodium chloride) and an internal standard (eg, tertiary butanol) and maintained in a sealed vial under thermostatted conditions until equilibrium is reached between the liquid and gas phases. An aliquot of the phase above the liquid (head space) is injected through a gas chromatographic column containing an appropriate stationary phase such as carbowax.[6]

Many procedures are available for the analysis of barbiturates and drugs of abuse after appropriate sample treatment. In one procedure, the extraction of drugs from a serum sample is performed on a disposable column of kieselguhr in the presence of phosphate buffer containing an internal standard. After allowing three minutes for the drugs to be adsorbed on the column, methylene chloride is added to elute the drug. The eluate is evaporated to dryness, reconstituted with methylene chloride, and injected into a column containing 2% methyl phenyl silicone (SP-2110) as the stationary phase.[8] The stationary phase (3% OV-17 methyl phenyl silicone) used with the nitrogen phosphorus detector has a wide variety of applications in the analysis of tricyclic antidepressants and selected antiarrhythmic drugs.[9–11] The application of gas chromatography has also extended to bacteriological assays for identification of microorganisms by studying their cellular fatty acid profile, bacterial metabolic products, and infection-induced substances in biological fluids (eg, lactate concentration in cerebrospinal fluid for rapid diagnosis of meningitis).[10–13] Capillary gas chromatographic procedures have found application in the measurement of catecholamines,[14] polyamines,[15,16] and organic acids in urine,[17] and in the measurement of volatile alcohols and acetone in serum.[18,19] Gas chromatographic procedures have been utilized for the measurement of gases such as carbon monoxide in blood[20] and hydrogen in breath samples.[21]

Recent developments in automation will increase the applications for gas chromatography because sample injection steps, including the use of innovative solid injection techniques, can be automated. In addition, setting analytical parameters such as carrier gas flow rates has been simplified with the use of digital switches.[22]

References

1. McNair HM, Bonelli EJ: Basic Gas Chromatography, ed 5. Walnut Creek, Calif. Varian Aerograph, 1969, pp 1–303.
2. Willard HH, Merritt LL Jr, Dean TA, Settle FA: Instrumental Methods of Analysis, ed 6. New York, Van Nostrand, 1981, pp 454–494.
3. Vanderborgh NE, Ristan WT: Laser pyrolysis gas chromatography for the characterization of solid materials. In Laboratory Instrumentation, series II, vol 1. Fairfield, Conn, International Scientific Communications, 1977, pp 28–32.
4. Horvath, C, Ettre LS, Purcell JE: Support-coated open tubular columns. In Laboratory Instrumentation, series II, vol 1. Fairfield, Conn, International Scientific Communications, 1977, pp 41–50.
5. Hendrickson JL, Wallace DL: Design of a microprocessor based data system for chromatography. In Laboratory Instrumentation, series II, vol 1. Fairfield, Conn, International Scientific Communications, 1977, pp 123–128.
6. Ettre LS, Kolb B, Hurt SG: Techniques of head space gas chromatography. American Laboratory 15:76, 1983.
7. Penton Z: Headspace measurement of ethanol in blood by gas chromatography with a modified autosampler. Clin Chem 31:439, 1985.
8. Werner M, Mohrbacher RJ, Riendeau CJ: Gas-liquid chromatography of underivatized drugs after chromatographic extraction from blood. Clin Chem 25:2020, 1979.
9. Bailey DN, Jatlow PI: Gas-chromatographic analysis for therapeutic concentrations of amitriptyline and nortriptyline in plasma, with use of a nitrogen detector. Clin Chem 22:777, 1976.
10. Vinet B: Column deactivation in analysis for underivatized tricyclic antidepressants by gas chromatography with use of a nitrogen detector. Clin Chem 29:452, 1983.
11. Kessler KM, Ho-tung P, Steele B, et al: Simultaneous quantification of quinidine, procainamide and N-acetylprocainamide in serum by gas-liquid chromatography with a nitrogen-phosphorus selective detector. Clin Chem 28:1187, 1982.
12. Lambert MS, Armfield AY: Differentiation of peptococcus and peptostreptococcus by gas-liquid chromatography of cellular fatty acids and metabolic products. J Clin Microbiol 10:464, 1979.
13. Gastrin B, Briem H, Rombo L: Rapid diagnosis of meningitis with use of selected clinical data and gas-liquid chromatographic determination of lactate concentration in cerebrospinal fluid. J Infect Dis 139:529, 1979.
14. Tuchman M, Crippin PJ, Krivit W: Capillary gas-chromatographic determination of urinary homovanillic acid and vanillmandelic acid. Clin Chem 29:828, 1983.

15. Muskiet, FAJ, Van den Berg GA, Kingma AW, Fremouw-Ottevangers DC, Halie MR: Total polyamines and their non-α-amino acid metabolites simultaneously determined in urine by capillary gas chromatography, with nitrogen-phosphorus detector; and some clinical applications. Clin Chem 30:687, 1984.

16. Van den Berg GA, Muskiet FAJ, Kingma AW, Van der Slik W, Halie MR: Simultaneous gas-chromatographic determination of free and acetyl-conjugated polyamines in urine. Clin Chem 32:1930, 1986.

17. Chamberlin BA, Sweeley CC: Metabolic profiles of urinary organic acids recovered from absorbent filter paper. Clin Chem 33:572, 1987.

18. Smith NB: Determination of volatile alcohols and acetone in serum by non-polar capillary gas chromatography after direct sample injection. Clin Chem 30:1672, 1984.

19. Penton Z: Gas-chromatographic determination of ethanol in blood with 0.53-mm fused-silica open tubular columns. Clin Chem 33:2094, 1987.

20. Vreman HJ, Kwong LK, Stevenson DK: Carbon monoxide in blood: an improved microliter blood sample collection system, with rapid analysis by gas chromatography. Clin Chem 30:1382, 1984.

21. Gaffney PT, Buttenshaw RL, Thomas MJ, Read RSD, Ward M: Faster assay of H_2 in breath by dedicated instruments compared with conventional gas chromatography. Clin Chem 32:1784, 1986.

22. Kahn HI, Gill JM: Trends in chromatography automation. In Laboratory Instrumentation, series II, vol 1. Fairfield, Conn, International Scientific Communications, 1977, pp 139–142.

15
Mass Spectrometry

Mass spectrometry in the clinical laboratory is used as a valuable adjunct to identification by gas chromatography. Although gas chromatography identifies a constituent by its retention time, it is quite possible that a closely related constituent, having a similar affinity for the stationary phase of the column, could be eluted with the same retention time, and thus be misidentified. However, mass spectrometric analysis of the eluted peak from the gas chromatograph permits the determination of the molecular weight of the compound, and even the structure based on the fragmentation patterns of the molecule. It thus affords positive identification. Mass spectrometric analysis by itself is a powerful tool in the determination of molecular weight and structure of unknown constituents in a biological sample.

When a vaporized sample molecule is bombarded with electrons, an electron can be removed from the molecule with the formation of a positively charged molecular or parent ion. The mass of this parent ion is equal to the molecular weight of the compound because loss of an electron does not affect its mass, the mass of the electron being infinitesimally small. The process can be expressed by the following equation:

$$\underset{\text{Molecule}}{R{:}R^1} \xrightarrow{e^-} R^{\dot{+}}\, R^1 + 2e^-$$

Many of these parent ions disintegrate rapidly between 10^{-10} to 10^{-8} seconds resulting in the formation of a positively charged fragment ion and a radical:

$$(R^+\, R' \longrightarrow R^+ + \cdot R^1)$$

Parent ions that survive without undergoing fragmentation and that reach the detector in about 10^{-6} seconds provide a parent peak that allows the determination of the molecular weight of the compound.

Depending on the intensity of the electron beam, a number of fragment ions are produced that provide information about the structure of the compound. A mass spectrum is a plot of masses of positively charged fragments [mass/charge (M/e) ratio] including the parent ion vs their relative intensity. The peak that has the greatest intensity is designated as the base peak, and is assigned an intensity of 100%. The intensities of other peaks are expressed as percentages of the base peak. Peaks due to isotopes can also be noticed on the mass spectrum. Knowing the percentage of a relative abundance of isotopes (which can be determined on the basis that the isotope of the lowest mass number is 100%) permits the identification of compound from a list of compounds having the same molecular weight.

Design and Components

A mass spectrometer consists of: a sample inlet where sample is vaporized under a high vacuum; an ion source to ionize and hold the sample for approximately a microsecond to favor fragmentation of the sample; a device called the mass analyzer or mass filter to separate the molecular ions and fragment ions according to their mass/charge (M/e ratio); a detector to detect the relative intensity or abundance of these masses; an amplifier to amplify the detector current; and a recorder to print out the mass spectrum.[1] Mass spectrometers have computerized capabilities to facilitate data processing, selective ion monitoring in the picogram range, and matching up with reference spectra to confirm identity of unknown compound.

Sample Inlet

The sample is vaporized at the inlet, which is under a high vacuum with pressure at the inlet as low as 10^{-7} to 10^{-6} torr (1 torr = 1 mmHg). The sample inlet is also maintained at a high temperature of approximately 300°C to insure maintenance of the sample in the gaseous state and to avoid condensation of sample that would otherwise contribute to background noise and result in carryover into the succeeding samples. Low pressures of 10^{-7} torr are achieved through the use of high speed vacuum pumps that can pump 700 to 1000 L/sec. If low pressures of the order of 10^{-6} to 10^{-7} torr are not maintained, the resulting excess of background air molecules will collide with ionized masses generated in the ion source, thus interfering with their flight towards the detector.

Volatile samples or samples rendered volatile by derivatization when introduced into the sample inlet are vaporized and subsequently enter the ion source. Nonvolatile or thermally labile samples can, however, be directly introduced into the ion source.

Ion Source

The function of the ion source is to ionize the sample. Depending on the type of ionization technique used, one may obtain a spectrum simple enough to identify the molecular or parent ion. One can then ascertain the molecular weight of the compound, or a complex spectrum in which the molecule is broken literally into many fragments of varying masses. Analysis of these masses permits determination of the structure of the compound.

Among the ionization techniques available, the electron-impact technique permits obtaining a detailed fragmentation pattern. Simpler spectra are obtained with chemical ionization, field ionization, or field desorption techniques.

Electron impact. The vaporized sample entering through the sample inlet is ionized by this ion source. The source of electrons is a glowing filament made of either tungsten, rhenium, or iridium. The filament is placed at a negative potential of 70 V. The resulting electron beam has a very high kinetic energy of approximately 70 eV, which is sufficient to remove an electron from most molecules (whose ionization potentials usually range from 8 to 13 eV) to produce a molecular ion. Some of the molecular ions are in a higher energy state and fragment further to produce a detailed mass spectrum. It is possible to control the kinetic energy produced by the electron source by varying the voltage applied to the filament. Thus, at low voltage, one can produce electron beams of 12 to 15 eV that can be used to produce a simpler spectrum.

The vaporized molecules entering the ion source that escape ionization are pumped away, and the positively charged molecular ions or ionic fragments are accelerated under applied voltage (1,000 to 10,000 V) to the mass analyzer or mass filter portion of the spectrometer for further manipulation. Although a small number of negative ions are also produced during ionization, which may be as low as 1 negative ion for every 100 positive ions that result from ionization, the mass spectrometer is designed to measure positively charged species.

The flight of the positively charged molecular ions and ionic fragments toward the mass analyzer is facilitated by placing in the analyzer's path in the ion source repeller plates, which are at a slightly higher positive potential than the ion source. The repelled positively charged ions and ion fragments are drawn toward negatively charged accelerator plates positioned near the entrance of the mass analyzer portion of the instrument, into which they are readily focused with the aid of focusing plates.

Chemical ionization. In this technique, a gas such as methane is admitted into the ion source and ionized by electron-impact. The entrance of gas into the ion source raises the pressure to as high as 0.1 torr, favoring the reaction of ionized methane gas with nonionized methane gas to produce a reactive intermediate CH_5+, which in turn transfers a positive charge or proton to the sample molecule to produce an M+1 ion. The molecular weight of this $(M+1)^+$ ion is equal to the molecular weight of the compound plus one, since a proton has been added during charge transfer. The events that take place during chemical ionization can be summed up by the following sequence of reactions.

$$CH_4^{+} + \xrightarrow{e^-} CH_4^{+}\cdot + 2e-$$

(Methane gas)

$$CH_4^{+}\cdot + CH_4 \longrightarrow \cdot CH_3 + CH_5^{+}$$

Other unionized methane gas

$$CH_5^+ + M \rightarrow M + H^+ + CH_4$$

(Sample Molecule) $(M + 1)^+$

Molecule + H^+

The $(M + 1)^+$ species can be fragmented further. In general, the chemical ionization technique that produces $(M + 1)^+$ ions and few fragment ions permits the determination of the molecular weight of the sample.

One can, with oxygen or hydrogen as an ionizing gas, produce negative ion $(M-1)$. The negative chemical ionization technique is especially suited to the detection of molecules that have a halogen function, such as chloride or fluoride.

Field ionization. In field ionization, vaporized molecules emerging from the sample inlet are ionized in a strong electrical field of approximately 10^7 V.CM^{-1}. In this technique, stable molecular ions are produced, which allows determination of the molecular weight of the compound. This technique, however, is less sensitive than electron impact or chemical ionization.

Field desorption. In this technique, the sample is deposited or adsorbed on a heated filament, and subsequently ionized by application of a positive potential. In this technique, the predominant species produced is the molecular ion, which permits the determination of the molecular weight of the compound. The spectrum obtained by this technique is simpler than that obtained by chemical ionization, because very few fragment ions are formed. The technique is suited to the determination of molecular weight of pure nonvolatile compounds that are not readily ionized by other techniques.

Mass Analyzer and Mass Filter

Molecular ions or ion fragments emerging from the ion source need to be sorted out according to their mass to charge (M/e) ratios. This can be accomplished by either using a magnetic field (the magnetic sector mass analyzer), or an electrical and radiofrequency field (the quadrupole mass filter).

In the magnetic sector mass analyzer, a magnetic field is used to deflect the ions according to their mass to charge (M/e) ratio. The molecular ions and ion fragments from the ion source are led through a curved tube, which is placed between the poles of a magnet, the magnetic sector. It is here that the ions are deflected according to their masses. The heavier ions are deflected the least, while the lighter ions are deflected the most. The radius of deflection necessary for an ion of proper charge and mass to reach the detector is determined by the curvature of the tube. The masses of the ions and ion fragments can be sorted out by keeping the magnetic field constant and decreasing the accelerating voltage. The masses are then deflected according to their increasing weights, so that each of the ions and ion fragments are brought into focus at the detector when the radius of deflection of the ion mass becomes equal to the radius of the curvature of the tube. Alternatively, as is the case with many of the current magnetic sector instruments,

Figure 15–1

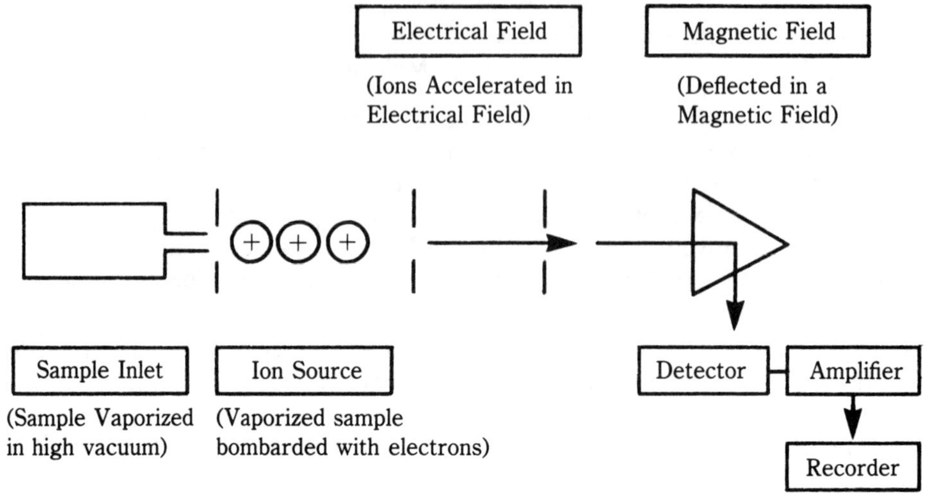

Simplified design of a magnetic sector mass spectrometer.

the magnetic field is varied at constant voltage to bring the various masses into focus at the detector.

Magnetic sector mass analyzers can be made to provide very high resolution by incorporating, prior to the magnetic sector, two electrodes that function as an electrostatic analyzer serving to sort out ions of equal kinetic energy. Because of the addition of the electrostatic analyzer, these mass analyzers are termed double-focusing mass spectrometers, as opposed to single-focusing mass analyzers that have only the magnetic sector. In short, double-focusing mass spectrometers separate ions according to their kinetic energy, and mass selection is performed in the usual manner by magnet. The accuracy of the determination of the molecular weight of a compound is dependent on the resolution of the instrument. High-resolution double-focusing mass spectrometers are able to compute the molecular weight of a compound accurate to 10^{-4} atomic mass units. Magnetic sector instruments can measure masses of the order of 2000 atomic mass units (amu).

Figure 15–1 represents a simplified design of a magnetic sector mass spectrometer.

In the quadrupole mass filter, the molecular ions and ion fragments emerging from the ion source are led through four metallic electrodes. A combination of DC voltage and radio frequency voltage produces an oscillating electrical field between the four electrodes (quadrupoles). Only certain ions take the flight path to the detector and are detected. These ions are considered to be in resonance. Masses that are too light or too heavy take contorted paths and miss the flight path to the detector, and are literally filtered out by collision with the electrodes. By varying the frequency of the radio frequency field, it is possible to bring the various ionized masses into focus or resonance at the detector, and thus be able to rapidly determine the mass spectrum of a compound. Quadrupole mass filters are used with gas chromatography–mass spectrometry (GCMS) interfaces since rapid scanning can be accomplished in less than 1 second. Masses up to 1000 amu can be scanned by these quadruple mass filters.

Detector

The detector used in mass spectrometers is an electron multiplier. It is similar to the photomultiplier tube used in spectrophotometry because the positive ion masses striking the first dynode produce electrons that are amplified at succeeding dynodes to produce a millionfold amplification of current. The amplified current can be fed to a strip-chart recorder. Spectra can be recorded by recorders with five separate galvanometers that permit simultaneous recording at five different sensitivity levels on either photographic paper or UV recording paper that does not require wet development. Sensitivity levels in the five galvanometer spectra decrease from top to bottom in the ratio of 1:3:10:30:100. Peak heights of masses are computed from the baseline on the most sensitive tracing on scale and are adjusted by multiplication by the appropriate sensitivity factor. Peak heights are proportional to the number of ions of each mass measured by the mass spectrometer.

Computerized data systems can handle all of the data reduction in tabular form or as a spectrum. They can print out the entire spectrum at command, and the total ion current chromatogram or selective ions can be monitored (selective ion chromatogram). Computerized systems are able to compare unknown spectra with a library of reference spectra to correctly identify the unknown sample.

Gas Chromatography—Mass Spectrometry Interface

Gas chromatographs operate at pressures just greater than atmospheric pressure (> 760 torr). In contrast, mass spectrometers operate at pressures in the range of 10^{-7} to 10^{-6} torr. The use of high-speed vacuum pumps in mass spectrometry ensures maintenance of high vacuum, which can be affected by a high flow rate of carrier gas from a gas chromatographic column. For this reason, most of the carrier gas from a conventional packed gas chromatographic column with flow rate approximating 40 mL/min is pumped away through a separator interface prior to the introduction of sample constituents into the mass spectrometer ion source. Either silicone rubber or teflon membrane separator, or fritted glass separator or jet separator have been used to remove carrier gas from the gas chromatographic effluent. The differential solubility of sample constituents in the lipophilic membrane of the membrane separator serves to enrich these sample constituents prior to their introduction into the ion source of the mass spectrometer, whereas the lighter carrier gas molecules that are not adsorbed by the membrane are readily pumped away by an external vacuum pump. In the fritted glass separator, the lighter carrier gas molecules preferentially diffuse and are pumped out, while vaporized sample constituents enter the ion source for ionization and analysis. The jet separator utilizes a narrow inlet through which the effluent from the gas chromatograph is introduced. The lighter carrier gas molecules diffuse rapidly and are pumped away while the sample molecules are led through a narrow outlet into the ion source. The gas chromatographic effluent inlet interface with the mass spectrometer is heated to the same temperature as the mass spectrometer to avoid condensation of sample constituents.

With the increasing use of capillary columns in gas chromatography, it has become possible to dispense with the separator device altogether, because the capillary column effluent can be introduced directly into the ion source. This is possible because the low flow rate of carrier gas through capillary columns is in the range of 1 to 4 mL/min, which can conveniently be handled by high-speed vacuum pumps used to maintain the very low pressures required in mass spectrometry.

Applications

Capillary gas chromatography-mass spectrometry (GCMS) has been widely used to unambiguously confirm the identity of peaks obtained on a gas chromatogram. It has been very valuable in the clinical toxicology laboratory. Quantitative GCMS procedures are available for quantification of hypnotic sedatives in serum.[2] The availability of fused-silica capillary columns in GC permits analysis of underivatized drugs and subsequent detection by MS. Thus procedures have been described for the qualitative analysis of neutral and basic drugs,[3] and acidic and basic drugs.[4] The availability of computer-operated bench-top capillary GCMS instruments for use in the clinical laboratory have further facilitated identification of drugs in clinical specimens such as blood and urine.[5] The availability of liquid chromatography mass spectrometry (LCMS) interface has also facilitated drug analysis.[6,7] GCMS has found application in the analysis of volatile compounds in saliva. Headspace analysis of saliva by GCMS has demonstrated differences in the nature of volatile compounds found in the saliva of healthy subjects and those with periodontal disease. In the saliva of the latter group, the presence of volatile aromatic amines such as pyridine and picolines was demonstrated, whereas these volatile amines were absent in the saliva of healthy subjects.[8]

Computer assisted GCMS procedures have been used in breath analysis and to detect differences in volatile compounds in the exhaled air of healthy and diseased subjects. Using this technique, it was demonstrated that the exhaled air of a majority of lung cancer patients contained four volatile constituents that were absent from the exhaled air of healthy subjects.[9]

The introduction of tandem mass spectrometry (MS/MS), which literally dispenses with the need to have a preliminary GC or HPLC column, confers picogram to femtogram sensitivity to breath analysis. In such tandem mass spectrometry systems (MS/MS), separation is effected in a mass spectrometer. The molecular ions resulting in the mass spectrometer are fragmented further and are separated by a second mass spectrometer according to their mass-to-charge ratio.[10] Breath analysis by direct-sampling atmospheric pressure ionization coupled to tandem mass spectrometry has great potential in breath analysis and other clinical assays[11] and is currently limited only by the relatively high cost of such instrumentation.

A technique that relies on the ionization of sample (dissolved in a liquid matrix such as glycerol) when it is bombarded by a beam of neutral atoms such as argon or xenon or positively charged cesium ion beam is called fast atom bombardment (FAB).[12,13] Positive and negative ion mass spectra are obtained with this technique, which is also referred to as secondary ion mass spectrometry (SIMS). Relatively simple spectra are obtained for most molecules with very little fragmentation. The FAB mass-spectrometric technique is valuable in the characterization of polar charged molecules. It has been used in the detection of steroid conjugates (glucuronides and sulfates) found in the urine of patients with inborn errors of steroid biosynthesis caused by specific enzyme defects.[14] It is also used for the measurement of urinary pregnanediol-3 α-glucuronide for assessing luteal function.[15]

Metabolic profiling by assay of urinary carboxylic acids have been performed by mass spectrometry.[16] Volatile metabolites in urine of diabetic subjects have been studied with GCMS to detect abnormalities associated with diabetes.[17]

Molecules in the molecular weight range of 500 to 2000, the so-called middle molecules, found in the serum of uremic subjects, have been the subject of study by GCMS.[18]

Among innumerable applications of GCMS, mention should be made of the assay for vitamin D in plasma,[19] testosterone in plasma from men,[20] and assay of melatonin metabolites in urine to understand the physiological role of the pineal gland.[21]

Finally, external quality control data have been validated by target values set by GCMS.[22,23] The accuracy of isotope dilution–mass spectroscopy has lent itself to the establishment of definitive methods for clinical chemistry analytes.[24,25].

References

1. Willard HH, Merritt LL Jr, Dean JA, Settle FA Jr: Instrumental Methods of Analysis, ed 6. New York, Van Nostrand, 1981, p 565.
2. Soo VA, Bergert RJ, Deutsch DG: Screening and quantification of hypnotic sedatives in serum by capillary gas chromatography with a nitrogen-phosphorous detector, and confirmation by gas chromatography-mass spectrometry. Clin Chem 32:325, 1986.
3. Cailleux A, Turcant A, Premel-Cabic A, Allain P: Identification and quantitation of neutral and basic drugs in blood by gas chromatography and mass spectrometry. J Chromatogr Sci 19:163, 1981.
4. Ehresman DJ, Price SM: Screening biological samples for underivatized drugs using a spitless injection technique on fused silica capillary column gas chromatography. J Anal Toxicol 9:55, 1985.
5. Deutsch DG, Bergert RJ: Evaluation of a bench top capillary gas chromatograph-mass spectrometer for clinical toxicology. Clin Chem 31:741, 1985.
6. Eckers C, Skrabalak DS, Henion J: On-line direct liquid introduction interface for micro-liquid chromatography/mass spectrometry: Application to drug analysis. Clin Chem 28:1882, 1982.
7. Lien EA, Ueland PM, Solheim E, Kvinnsland S: Determination of tamoxifen and four metabolites in serum by low-dispersion liquid chromatography. Clin Chem 33:1608, 1987.

8. Kosteic JG, Zelson PR, Preti G, Tonzetich J: Quantitative differences in volatiles from healthy mouths and mouths with periodontitis. Clin Chem 27:842, 1981.
9. Gordon SM, Szidon JP, Krotoszynski BK, Gibbons RD, O'Neill HJ: Volatile organic compounds in exhaled air from patients with lung cancer. Clin Chem 31:1278, 1985.
10. McLafferty FW: Tandem mass spectrometry. Science 214:280, 1981.
11. Manolis, A: The diagnostic potential of breath analysis. Clin Chem 29:5, 1983.
12. Barber M, Bordoli RS, Sedgwick RD, Taylor AN: Fast atom bombardment of solids as an ion source in mass spectrophotometry. Nature 293:270, 1981.
13. Aberth W, Straub KM, Burlingame AL: Secondary ion mass spectrometry with cesium ion primary beam and liquid target matrix for analysis of bio-organic compounds. Anal Chem 54:2029, 1982.
14. Shackleton CHL: Inborn errors of steroid biosynthesis: Detection by a new mass-spectrometric method. Clin Chem 29:246, 1983.
15. Monet G, Agati G, Giovannini MG, et al: Pregnanediol-3α-glucuronide measured in diluted urine by mass spectrometry with fast atom bombardment/negative-ion ionization. Clin Chem 31:46, 1985.
16. Hunt DF, Giordani AB, Rhodes G, Herold DA: Mixture analysis by triple-quadrupole mass spectrometry: Metabolic profiling of urinary carboxylic acids. Clin Chem 28:2387, 1982.
17. Rhodes G, Miller M, McConnell ML, Novotny M: Metabolic abnormalities associated with diabetes mellitus, as investigated by gas chromatography and pattern-recognition analysis of profiles of volatile metabolites. Clin Chem 27:580, 1981.
18. Schoots AC, Mikkers FEP, Claessens HK, et al: Characterization of uremic "middle molecular" fractions by gas chromatography, mass spectrometry, isotachophoresis, and liquid chromatography. Clin Chem 28:45, 1982.
19. Coldwell RD, Trafford DJH, Makin HLJ, Varley MJ, Kirk DN: Specific estimation of 24, 25-dihydroxy vitamin D in plasma by gas chromatography-mass spectrometry. Clin Chem 30:1193, 1984.
20. Finlay EMH, Gaskell SJ: Determination of testosterone in plasma from men by gas chromatography/mass spectrometry, with high-resolution selected-ion monitoring and metastable peak monitoring. Clin Chem 27:1165, 1981.
21. Francis PL, Leone AM, Young IM, Stovell P, Silman RE: Gas chromatographic-mass spectrometric assay for 6-hydroxymelatonin sulfate and 6-hydroxymelatonin glucuronide in urine. Clin Chem 33:453, 1987.
22. Gaskell SJ, Collins CJ, Thorne GC, Groom GV: External quality assessment of assays for cortisol in plasma: Use of target data obtained by gas chromatography/mass spectrometry. Clin Chem 29:862, 1983.
23. Gaskell SJ, Sieckmann L: Mass-spectrophotometric determination of serum cortisol: Comparison of data from two independent laboratories. Clin Chem 32:536, 1986.
24. Pelletier O, Wright LA, Breckenridge WC: Isotope dilution/mass spectrometry of serum cholesterol with [3,4-^{13}C] cholesterol: proposed definitive method. Clin Chem 33:1403, 1987.
25. Pelletier O, Arratoon C: Precision of glucose measurements in control sera by isotope dilution/mass spectrometry: Proposed definitive method compared with a reference method. Clin Chem 33:1397, 1987.

16
High-Performance Liquid Chromatography

The "high performance" of HPLC is achieved by a considerable increase in column efficiency, made possible by the use of packing materials of small particle size (5 to 20 μm). The considerable resistance created by these packing materials to flow of solvent through the column is overcome by the high pressures used to pump solvent through the column. Indeed, HPLC stood for high-*pressure* liquid chromatography in the early days of the technique. Because there are variables other than pressure that determine optimal performance, however, high *performance* liquid chromatography best describes this technique.

Design and Components

Figure 16–1 depicts the components that generally make up an HPLC system: solvent reservoirs; gradient mixers to mix solvents in a definite ratio achieved through proportioning valves that control flow of solvent from reservoirs; pumps and pressure regulators to pump solvents through the column at high pressure; a filter to remove particles in the degassed solvents; a precolumn to improve the performance of the analytical column; a sample injection device to introduce sample into the main analytical column or into a guard column connected to the analytical column in order to extend its life; the main analytical column (enclosed in a thermostatted chamber) containing the packing material; the detector to monitor the constituents emerging from the column; and, optionally, a fraction collector used to collect sample constituents for further analysis for verification of identity.

Solvents or Mobile Phase

Choice of solvents for the mobile phase is based on several considerations, such as viscosity, compatibility with the detector used, polarity, selectivity, compatibility

Figure 16–1

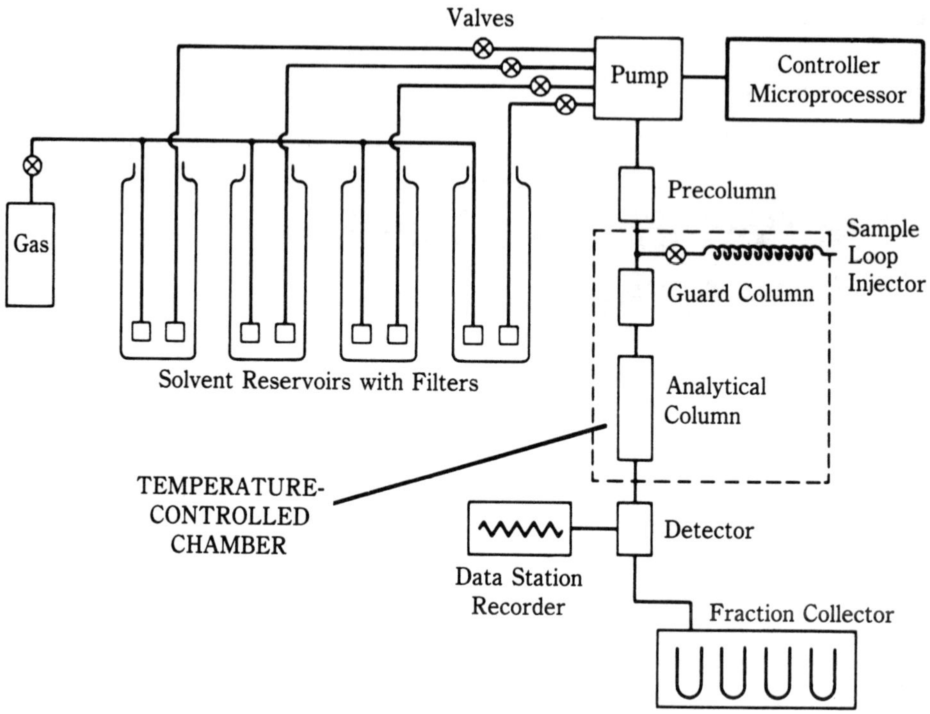

Components of an HPLC system.

with the column stationary phase, nature of sample constituents, type of chromatography, and safety.

In general, increasing solvent viscosity decreases column efficiency and flow rate and prolongs analysis time. To maintain a uniform flow rate, pump pressure has to be increased in direct proportion to the increment in solvent viscosity.[1]

The solvents should be compatible with the detector used for monitoring sample constituents. Thus, solvents should have a low refractive index, especially when used with refractive index detectors, since the sensitivity of this detector is directly related to the difference in refractive index of the solvents and sample constituents.[2] Ideally, the mobile phase solvent should have negligible absorbance at the wavelength of absorbance measurements of sample constituents. In practice, however, olefinic impurities or antioxidant stabilizers, such as butylated hydroxytoluene, added to ethers to prevent formation of explosive peroxides, may absorb in the ultraviolet range. For this reason, it is important to use only high-purity solvents or to subject solvents to a purification step immediately prior to use.

Solvent polarity is dependent on the strength of attraction between sample and solvent molecules. The extent of such attraction is dependent on the dipole moments of solvent and sample, the dielectric constant of the solvent, the magnitude of the refractive index of solvent and sample giving rise to dispersion interactions, and the extent of hydrogen bonding. Because polar sample constituents dissolve in polar solvents, the ability of solvents to preferentially elute polar sample constituents in partition liquid chromatography is directly related to solvent strength, which in turn is a reflection of solvent polarity. The choice of solvents based on their strength or polarity has been simplified by the availability of tables listing solvents on the basis of their polarity index (P^1) developed from solubility data.[3]

P^1 is a sum of three component interactions: dipole (χ_n), acid or proton donor (χ_d), and base or proton acceptor (χ_e). Choice of a solvent mixture to accentuate any of these interactions achieves mobile phase selectivity. A change in solvent or mobile phase selectivity (α^1) is needed to separate overlapping sample constituent peaks in HPLC. Solvents have been classified into eight groups based on their selectivity. Group 1 solvents, such as ethers and amines, have increased dipole interaction and are good proton acceptors, but poor proton donors; in contrast, group 8 solvents, such as chloroform, are excellent proton donors, but poor proton acceptors and have weak dipole interactions. Thus, the appropriate choice of mobile phase solvent pairs based on solvent selectivity can resolve overlapping sample constituent peaks. For instance, the choice of a solvent with increased dipole interaction will facilitate retention by solvent of sample constituents with pronounced dipole moment. Similarly, a solvent that is a good proton donor will retain sample constituents that are good proton acceptors.

The pH of the mobile phase solvent mixture may have an effect on the column stationary phase. Above pH 7 to 8, silica gel is subject to attack and may begin to dissolve.[4] The base used to control pH of the mobile phase may have an effect on the stability of the stationary phase. Thus, the use of quaternary alkyl amines greatly affects the stability of bonded silica gels used in reverse-phase chromatography when compared with the use of tertiary amines. In contrast, even at pH 12.5 the use of triethylamine does not affect the stability of bonded silica gel stationary phase.[5]

The choice of mobile phase solvent mixtures depends not only on the chemical nature of the sample, but also on the type of HPLC.

Finally, the choice of a solvent is dictated by safety considerations. Toxicity guidelines provided by the Occupational Safety and Health Administration (OSHA) about threshold limit value (TLV), which stipulates maximum concentration of solvent vapor that can accumulate in the working area, should be consulted. Data on flashpoint of solvents, which indicates the temperatures at which the solvent vapor becomes flammable, should also be consulted. Because the solvents used in HPLC are relatively flammable, it is important that the work area be adequately ventilated.

Gradient Mixers

It may at times be necessary to vary the composition of the solvent mixture (gradient elution) as the chromatography proceeds. Gradient elution reduces analysis time compared with isocratic elution, in which the composition of the mobile phase is kept constant throughout the chromatographic run. Gradient elution also allows increasing selectivity by appropriate changes in solvent composition. In general, in gradient elution, constituents that elute early in isocratic elution are well separated and elution times of later-eluting constituents are shortened.

Gradient elution can be stepwise, with a single change from one solvent to another, or continuous, with composition varying throughout chromatography. Gradient mixers can provide gradients of various shapes (linear, convex, or concave). HPLC instruments currently available allow gradient elution functions to be programmed with the aid of a microprocessor.[6]

Pumps

The function of pumps is to ensure a constant, reproducible, pulse-free flow of mobile phase through the analytical column. They should be able to deliver solvents at pressures in the range of 500 to 5,000 psi. Pumps are usually constructed of materials resistant to chemical attack such as stainless steel or Teflon. Pump seals such as graphite-filled Teflon are also resistant to chemical attack. The stainless steel plumbing in pumps and columns can be attacked by the presence of halide impurities, such as HCl and HBr, which can be removed by passing the solvents through an alumina column. It is best to avoid mobile phases that are corrosive to stainless steel, such as mixtures of chloroform or carbon tetrachloride with ethers. If used, however, such mixtures should be prepared freshly prior to use.

Pumps that are used in HPLC deliver solvents either at a constant pressure (constant-pressure pumps) or at a constant flow rate (constant-displacement pumps). Pumps that deliver solvents at a constant flow rate, such as reciprocating pumps, are widely used in HPLC.[7]

Filter

Solvents deaerated either by refluxing and evacuating or heating, or by exposure to ultrasonic vibration, are filtered through a 0.5-μm membrane filter. The deaeration step ensures that the mobile phase is free of air bubbles as it passes through the detector, thereby eliminating an unstable baseline. It is also important to remove dissolved oxygen that might otherwise interact with the solvent or affect some HPLC determinations. Deoxygenation can be performed efficiently by bubbling helium through the solvent for approximately 30 minutes, a procedure called sparging.

HPLC instruments usually have either a 0.5-μm (especially with columns with particles 10 μm or smaller) or a 2-μm line filter to prevent any residual particulate matter in the solvent mobile phase from reaching the column.

Precolumn

A precolumn is sometimes used between the pump and the sample injector. This is especially useful to equilibrate, prior to flow into the analytical column, the mobile phase solvents that have greater than 50% mutual solubility.

A precolumn is also useful for minimizing dissolution of silica-based packing materials and, thus, prolonging the life of the analytical column. Such a column contains the same silica-based packing used in the analytical column. The effect of percolating the mobile phase through such a precolumn is to saturate the mobile phase with the silica stationary phase, thereby minimizing the dissolution of the stationary phase in the analytical column.

Guard Column

The purpose of a guard column inserted between the sample injector and the analytical column is to prolong the life of the analytical column. Packed with the same stationary phase used in the analytical column, the guard column serves to prevent impurities from reaching the analytical column. It also prevents strongly retained sample constituents from reaching the analytical column and affecting chromatographic resolution.

Sample Injector

Ideally, the sample injector should introduce the sample reproducibly as a narrow plug to ensure minimal peak broadening. The sample injector should also be able to operate at high backpressures.

Injection of sample directly into a pressurized column with the syringe through a self-sealing septum (syringe injection) has been largely supplanted by more reproducible injection with sampling valves. Sample valve injectors have a very high reproducibility and can operate at pressures up to 7,000 psi. The volume of sample introduced with sample valve injectors is dictated by either the size and internal diameter of the sample loop (fixed-loop injector), which is filled with a noncalibrated syringe, or by use of a calibrated syringe to inject a precisely determined volume into the sample loop (variable-volume injection). In fixed-loop injection, because the sample volume is dictated by the size and internal diameter of the sample loop, changing sample volumes requires using sample loops with appropriate sizes and internal diameter. The calibrated syringe used in variable-volume injection, however, provides the ability to vary with ease the volume of sample to be injected, although the precision of such an injector is dependent on the precision of the syringe.

There are available injectors that can introduce the sample at the center of the column head (point injection), thereby preventing the sample from reaching the walls of the column ("infinite diameter column") and yielding sharp and symmetrical peaks.

Automation of sample injection has been achieved with microprocessor-controlled automatic injectors that can sequentially inject several samples. Between each sample, the sample loop is flushed automatically, minimizing carryover.[7]

Columns

Columns used to pack the stationary phase are usually made of stainless steel that can withstand pressures up to 10,000 psi and vary from 10 to 150 cm in length and 2 to 5 mm in internal diameter. Column end fittings and connectors used to connect the column to other modules of the instrument should have minimum dead volume and ensure minimal peak broadening. Porous stainless steel frits or stainless steel screens are used at the end of the column to hold the sample packing. The pore size of these frits should be approximately one-fourth or one-fifth the diameter of the column packing, allowing uniform flow and minimizing peak broadening.

The use of flexible plastic-walled columns containing stationary phase permits radial compression of the column during chromatography.[8]

Column Packing Materials

Choice of column packing materials is made on the basis of particle size, surface area, and chemical characteristics. The smaller the particle size, the greater the column efficiency (an increase in the number of theoretical plates), which allows use of columns of shorter lengths. Thus, column packings of particle size 10 μm or smaller are increasingly being used in HPLC.

The surface area of the column packing is dependent on whether the particles are fully porous, superficially porous, or pellicular. Conventional fully porous particles (50 μm) have a large surface area, allowing sample molecules to penetrate deep into stationary phase and causing peak broadening because of stationary phase mass transfer effects. However, very small (5 to 10 μm) fully porous particles have negligible mass transfer effects and increased efficiency.

The superficially porous particle, however, has a relatively reduced surface area since the porous layer is built around a solid glass-bead core. This porous layer can either be reacted to yield a bonded-phase or be coated with a stationary phase. Porous or superficially porous column packings of particle size 5μm to 10μm or smaller have increased efficiency and sensitivity for trace analysis and provide rapid separation.

The pellicular packing has the least surface area since the stationary phase is deposited as a thin layer on a solid glass-bead core. However, because of the small quantity of stationary phase present in the pellicular packing, sample size becomes a limiting factor to prevent column overloading. In spite of this limitation, pellicular particles are relatively more stable, less expensive, and easy to pack, and are widely used in guard columns.

The stationary phase used in HPLC is usually covalently bonded to the surface of a silica particle. Nonpolar groups such as octyl silane (C_8) or octadecyl silane (C_{18}) and polar groups such as aminopropyl or cyanopropyl silanes can be bonded to the silica particle. The silanol (-Si-OH) group on the silica particle is reacted with a silylating agent bearing the stationary phase group. Thus, octadecyl dimethylchlorosilane is used to attach the octadecyl group to silica by converting a silanol group to a siloxane group (-Si-O-Si-). The resulting stationary phase is referred to as monomeric, since each molecule of silylating agent can react with only one silanol group. Approximately one-third of the silanol groups are converted to siloxane groups. The remaining unreacted silanol groups are reactive and may interact with polar sample constituents and affect the selectivity of the bonded nonpolar stationary phase. This can be overcome by another silanization reaction with trimethylchlorosilane or hexamethyldisilazane to convert the most accessible silanol groups to siloxane groups, a process called "capping off' or "end capping."[9]

Detectors

Detectors used in HPLC rely either on changes in the bulk property of the mobile phase and sample constituents, such as a change in the refractive index, or on some characteristic property of the sample constituent (solute property), such as fluorescence, ultraviolet absorption, or electrochemical changes.

Bulk property detectors such as refractive index detectors have little clinical application because of their low sensitivity (1 mg/L), sensitivity of the mobile phase to changes in temperature, and incompatibility with gradient elution. As such, HPLC detectors used for clinical applications are the "solute property" type.

Absorbance-type detectors allow either isolation of a fixed wavelength (for example 254 nm or other wavelengths by choice of filters in the range of 280 nm to 660 nm), or variable wavelengths through use of monochromator, especially below 254 nm.

Because flow cell volumes should be low (5 to 10 μL) to prevent peak broadening caused by mixing within the cell, high-intensity light sources such as mercury vapor lamp, deuterium, or tungsten-halide lamps are used. Both fixed-wavelength and variable-wavelength detectors can be used with gradient elution. Fixed-wavelength detectors have detection limits of 10 ng/L because of lower detector noise levels than usually encountered with variable wavelength detectors. The latter have detection limits of 1 ng/L.

Fluorescence detectors have a greater sensitivity than absorbance detectors. The detection limit can be as high as 100 ng/L. The flow cell volumes are usually 20 μL and the detector can be used with gradient elution. Fluorescence detectors use either a xenon-arc lamp, deuterium lamp, or even a laser beam to excite sample constituents to fluorescence emission. Fluorescence detectors are suited to measure constituents that have a native fluorescence or those than can be reacted with derivatizing agents, such as fluorescamine, dansyl chloride, or O-phthalaldehyde to give fluorescent compounds. Fluorescence measurements are, however, sensitive to changes in mobile phase pH, presence of dissolved oxygen, and impurities that might quench fluorescence.

Electrometric detectors can be used to detect compounds that can undergo either oxidation or reduction at an electrode surface. A constant voltage is applied between a working electrode whose potential has been set vs a reference electrode. Analyte oxidized or reduced at the potential of the working electrode causes a change in the net electron flow measured. Since current response is measured at fixed voltage, these detectors are termed amperometric detectors. An amperometric detector consists of three electrodes: a working electrode, a reference electrode, and an auxiliary electrode. The auxiliary electrode serves to complete the circuit with the working electrode, either by providing electrons by oxidizing the supporting electrolyte when an analyte is reduced at the working electrode potential, or by pulling electrons when an analyte is oxidized at the working electrode. The materials with which working electrodes are made vary from a dropping mercury electrode for electrochemical reductions to solid glassy-carbon electrodes for electrochemical oxidations. The electrodes used have a potential range of -1.3 V (or even -2.0 V) for reduction to $+1.5$ V for oxidation. Interference with electrochemical reductions is due to dissolved oxygen and trace metals that may also be reduced. This interference can be minimized by deoxygenating the mobile phase and introducing a chelant such as EDTA into the mobile phase to remove contaminating trace metals. Amperometric detectors use very small flow cell volumes (1 μL) and have detection limits of 1 pg/L for oxidizing species and 1 ng/L for reducing species. These detectors, however, are not suitable for gradient elution.

In some instances, reaction is effected after the mobile phase eluant leaves the column (postcolumn derivatization). The problem with postcolumn derivatization is the time delay needed for reaction to be complete before detection. This delay can be effected by using short lengths of capillary tubing (open tubular reactors), column or tubing filled with inert material (packed-bed reactors), or tubing with air-segmented streams (segmented-stream reactors). The segmented-stream reactors may have a time delay as long as 20 minutes depending on the kinetics of the reaction. It is important that the time delay between postcolumn reaction and detection be kept low to prevent peak broadening.

Data Reduction

Calculations from the strip-chart recorder output from the detector can best be handled by automatic data reduction systems that compute concentrations of unknown from peak height or area of unknown, internal standard, and standard.

Fraction Collector

An optional module for HPLC is a fraction collector, which can be useful in preparative chromatography to collect pure fractions of sample constituents.

Types of Chromatography

Bonded stationary phases used in HPLC can be either polar or nonpolar relative to the mobile phase. If the stationary phase is more polar than the mobile phase, the technique is referred to as normal-phase chromatography (NPC). The stationary phases used in NPC can be classified on the basis of polarity. Examples of high-polarity NPC stationary phases are those that contain amino, alkylamine, or aminopropyl functional groups. Examples of functional groups with medium polarity are those that contain nitrile, alkylnitrile, or nitro groups, whereas ester, diol, and dimethylamino function represent weak-polarity stationary phases. The choice of stationary phase with appropriate polarity would depend on the polarity of the sample. Mobile phases used in NPC are nonpolar solvents, such as hydrocarbons modified by slight admixture of moderately polar solvents such as ethers, ethylacetate, or chlorinated hydrocarbons. A mixture of hydrocarbon and an alcohol such as isopropanol may be used, but retention of polar sample constituents to the stationary phase could be reduced with an increase in concentration of alcohol. In NPC, in general, nonpolar sample constituents are eluted first. Solvent strength in NPC is judged by the increase in polarity of the solvent. As such, an increase in solvent strength in NPC would lead to competition of the mobile phase with the stationary phase for polar sample constituents and, hence, would decrease the elution time of the latter. If the stationary phase is less polar than the mobile phase, the technique is called reversed-phase chromatography (RPC), which is by far currently the most widely used HPLC technique in the clinical laboratory. Nonpolar stationary phases such as octadecyl (C_{18}) or octyl (C_8) bonded silica are used in RPC. Since the mobile phase is more polar than the stationary phase, polar sample constituents are eluted first in RPC. Solvent strength in RPC increases with decrease in polarity of the solvent. Thus, solvent strength increases, from water to dimethyl sulfoxide, ethylene glycol, acetonitrile, methanol, acetone, dioxane, ethanol, tetrahydrofuran, and isopropanol, in that order. Water constitutes the predominant mobile phase in RPC. Miscible organic solvents are added to water to control both solvent strength and selectivity. Methanol, acetonitrile, and tetrahydrofuran have been widely used as organic solvents to modify the mobile phase. Binary solvent mixtures such as water-methanol or water-acetonitrile and ternary solvent mixtures such as water-methanol-dioxane have been used to achieve the desired solvent strength and selectivity. The mobile phase in RPC may also include small amounts of acids such as sulfuric acid or phosphoric acids to suppress ionization and improve separation by reducing peak band tailing. The mobile phase may also be supplemented with salts such as ammonium formate or ammonium carbonate to reduce band tailing. Apparently, the addition of salts minimizes the interaction of sample constituents with some of the still-exposed silanol groups on the silica stationary phase support. In general, the use of a buffered mobile phase serves to maintain pH constant and an ionic strength above 0.01.

The popularity of RPC stems from the fact that a single stationary phase column such as octadecyl silica is adequate for most separations. The column packing is reasonably stable (three to six months) provided that the pH of the mobile phase is kept between pH 2 and pH 8.5. Water, the major mobile phase constituent, is plentiful, pure in terms of virtual absence of interfering substances, and cheap. Most separations can be performed without the use of highly flammable and relatively more toxic organic solvents. The column equilibrates rapidly to changes in mobile phase composition, and thus is very amenable to gradient elution. The versatility of RPC also lies in the facile manipulation of the mobile phase composition to effect the derived selectivity. Thus, the elution times of strongly retained nonpolar sample constituents can be reduced by decreasing the polarity of the mobile phase by increasing the concentration of the appropriate organic solvent. In general, a 10% increase in the organic solvent concentration in the aqueous mobile phase results in the reduction of K′ values two- to threefold. One of the limitations of RPC is that ionic samples will not be retained by the nonpolar stationary phase. This has been overcome by ionic suppression and ion pairing.

In ionic suppression, by buffering the mobile phase, the pH is maintained below the pK_a of the weak acid or weak base, so that they are unionized and are better retained by the nonpolar stationary phase. As was noted earlier, this technique also reduces peak band tailing with acids added to the mobile phase. The limitation of ionic suppression is that it is applicable only to weak acids and bases whose ionization can be suppressed in the pH range of 2 to 8. To depress the ionization of strong acids and bases would require pH change outside the range of stability of the bonded silica stationary phase (pH 2 to 8). Fortunately, ionic suppression of strong acids and bases has been addressed by a technique called ion-pair chromatography (IPC).

In IPC an organic counterion is used to form a neutral ion pair with the ionic species. The resulting ion pair, being neutral, its retained by the nonpolar stationary phase used in RPC, thus allowing good separation of ionized sample constituents.[10]

Essentially, IPC can be used to separate any ionic or ionizable species by RPC, although it is ideal for the separation of strongly ionized acids and bases.

The organic counterion that is added to the mobile phase (usually water-methanol) is of the opposite charge to that of the ionized sample constituent. Quaternary amines, such as tetramethyl or tetrabutyl ammonium chloride or hydroxide, are examples of counterions used to ion pair with acids (eg, sulfonic acids). Pentane or heptane sulfonic acids and perchlorates are examples of counterions used to ion pair with bases. In general, increasing the chain length of the counterion (pentane sulfonic acid vs heptane sulfonic acid or tetramethyl ammonium chloride vs tetrabutyl ammonium chloride) results in increased retention of the ion pair to the nonpolar stationary phase. The ion-pair technique has also been referred to as "soap" chromatography when detergents, which have good solubility in water-acetonitrile mobile phase, have been used as counterions. The concentration of counterions added to mobile phase in IPC is usually between 0.005 and 0.02 mol/L. IPC allows the separation of both polar and ionic sample constituents. However, the stability of the column packing is believed to be reduced by the addition of ion-pairing salts to the mobile phase.

Troubleshooting

The analyst usually encounters a host of practical problems, most of which can best be solved by consulting the manufacturer's preventive maintenance and troubleshooting manual. Some of the problems may relate to the recorder; for example, a noisy baseline may be due to several factors, including bubbles or particulate matter entering the flow cell or a leaking filter. Similarly, a drifting baseline could result from an overloaded column eluting strongly adsorbed constituents or from contamination in solvent reservoir. An increased sample retention time could be due to a slow flow rate or to a faulty gradient system. Two possible reasons for a loss in column resolution are overloading the column with the sample and a degradation of the column packing.

Applications

In addition to therapeutic drug monitoring, the area in which HPLC has found its widest application, HPLC has been successfully used in the assay of vitamins and selected hormones and their metabolites, and even analytes such as creatinine.

Antimicrobial agents such as aminoglycosides (streptomycin, kanamycin, gentamicin, tobramycin, amikacin), β-lactam antibiotics (penicillins, cephalosporins), bacteriostatic antibiotics (chloramphenicol, sulfonamides, trimethoprim), and broad-spectrum antibiotics (tetracyclines, vancomycin, lincomycin, clindamycin, erythromycin), and antifungal agents (griseofulvin, amphotericin B, flucytosine) have all been measured by HPLC.[11–15]

Several HPLC methods have been developed for the measurement of antiepileptic drugs such as barbiturates (phenobarbitol and primidone, which, although not a barbiturate, is closely related and is metabolized to the barbiturate phenobarbital), hydantoins (diphenylhydantoin), succinimides (ethosuximide), benzodiazepines (diazepam, chlordiazepoxide), and others (carbamazepine, valproic acid).[15-20] Antiarrhythmic agents (quinidine, lidocaine, procainamide, N-acetylprocainamide, propranolol), analgesics (salicylates, acetaminophen, phenacetin), bronchial smooth muscle relaxants (theophylline), and tricyclic antidepressants (amitryptyline, nortryptyline, imipramine, desipramine, doxepin, protriptyline) are the classes of drugs for which HPLC has found application.[15,21-23] The advantage of some of these HPLC methods is that a single chromatographic run can simultaneously resolve several drugs and some of their metabolites.

Other areas in which HPLC has found application include vitamins (vitamins A and E),[24-26] selected hormones (catecholamines,[27-31] cortisol,[32] estriol,[33]), proteins[34] and aminoacids,[35] creatinine,[36] uric acid,[37] glycosylated hemoglobin,[38] glycated proteins,[39] histamine,[40] and urine analysis of porphyrins,[41] carboxylic acids,[42] and glycosaminoglycans.[43]

The choice of HPLC procedure depends on the complexity of sample preparation steps. For instance, in the assay of plasma catecholamines, the acid extract of the plasma is absorbed onto, and subsequently eluted from, a weak cation-exchange resin prior to extraction with alumina.[18] Some constituents may require extraction with solid-phase adsorbent, elution, evaporation, and reconstitution with mobile phase prior to analysis. Assay for free drugs would require either ultrafiltration or dialysis. Some procedures require extraction with water-immiscible organic solvents and re-extraction of the desired analyte into the aqueous phase while rigidly controlling pH during extraction steps.

Frequently, however, the procedure may entail mixing the serum with mobile phase containing acetonitrile or methanol (both of which act as protein precipitating agents), centrifugation, and analysis of the supernatant containing the drug released from its protein-binding sites.

References

1. Bakalyar SR: Mobile phases for high performance liquid chromatography. Am Lab 10:43, 1978.
2. Snyder LR, Kirkland JJ: Introduction to Modern Liquid Chromatography, ed 2. New York, John Wiley & Sons, 1979, pp 246–268.
3. Snyder LR: Classification of the solvent properties of common liquids. J Chromatogr Sci 16:223, 1978.
4. Karger BL, LePage JN, Tanaka N: Secondary chemical equilibria in high-performance liquid chromatography. In Horvath C (ed): High-Performance Liquid Chromatography: Advances and Perspectives, vol 1. New York, Academic Press, 1980, p 133.
5. Wehrli A, Hildenbrand JC, Keller HP, et al: Influence of organic bases on the stability and separation properties of reversed-phase chemically bonded silica gels. J Chromatogr 149:199, 1978.
6. Snyder LR: Gradient elution. In Horvath C (ed): High-performance Liquid Chromatography: Advances and Perspectives, vol 1. New York, Academic Press, 1980, pp 207–316.

7. Snyder LR, Kirkland JJ: Introduction to Modern Liquid Chromatography, ed 2. New York, John Wiley & Sons, 1979, p 83–124.
8. Assenza SP, Brown PR: Evaluation of reversed-phase, radially-compressed flexible-walled columns for the separation of low molecular weight, UV-absorbing compounds in serum. J Liq Chromatogr 3:41, 1980.
9. Melander WR, Horvath C: Reversed-phase chromatography. In Horvath C (ed): High-Performance Liquid Chromatography: Advances and Perspectives, vol 2. New York, Academic Press, 1980, p 139.
10. Horvath C, Melander WR, Molnar I, et al: Enhancement of retention by ion-pair formation in liquid chromatography with non-polar stationary phases. Anal Chem 49:2295, 1977.
11. Cummings, KC, Torres AR, Edberg SC: The analysis of antimicrobial agents in biological fluids by high-performance liquid chromatography: II. Applications. J Clin Lab Autom 4:244, 1984.
12. McAteer JA, Hiltke MF, Silber M, Faulkner RD: Liquid-chromatographic determination of five orally active cephalosporins—cefixime, cefaclor, Cefadroxil, cephalexin, and cephradine—in human serum. Clin Chem 33:1788, 1987.
13. Lensmeyer GL, Wiebe DA, Carlson IH: Identification and analysis of nine metabolites of cyclosporine in whole blood by liquid chromatography. 1: Purification of analytical standards and optimization of the assay. Clin Chem 33:1841, 1987.
14. Lensmeyer GL, Wiebe DA, Carlson IH: Identification and analysis of nine metabolites of cyclosporine in whole blood by liquid chromatography. 2: Comparison of patients' results. Clin Chem 33:1851, 1987.
15. Gerson B, Anhalt JP: High-pressure liquid chromatography and therapeutic drug monitoring. Chicago, American Society of Clinical Pathologists, 1980, pp 81–263.
16. Kabra PM, Nelson MA, Marton LJ: Simultaneous very-fast liquid-chromatographic analysis of ethosuximide, primidone, phenobarbital, phenytoin, and carbamazepine in serum. Clin Chem 29:473, 1983.
17. Gerson B, Bell F, Chan S: Antiepileptic agents—primidone, phenobarbital, phenytoin, and carbamazepine by reversed-phase liquid chromatography. Clin Chem 30:105, 1984.
18. Matsumoto K, Kikuchi H, Kano S, et al: Automated determination of drugs in serum by liquid chromatography with column-switching. I: Separation of antiepileptic drugs and metabolites. Clin Chem 34:141, 1988.
19. Bock JL, Ben-Ezra J: Rapid measurement of free anticonvulsant drugs by direct liquid chromatography of serum ultrafiltrates. Clin Chem 31:1885, 1985.
20. Ray K, Trawick WG, Mullins RE: Assay of total and free propanolol in plasma by liquid chromatography with fluorometric detection. Clin Chem 31:131, 1985.
21. Kushida K, Oka K, Suganuma T, et al: Simultaneous determination of lidocaine and its principal metabolites by liquid chromatography on silica gel, with aqueous eluant. Clin Chem 30:637, 1984.
22. Sutfin TA, D'Ambrosio R, Jusko WJ: Liquid-chromatographic determination of eight tri- and tetracyclic antidepressants and their major active metabolites. Clin Chem 30:471, 1984.
23. Proeiss HF, Townsend TB: Simultaneous liquid-chromatographic determination of five antiarrhythmic drugs and their major active metabolites in serum. Clin Chem 32:1311, 1986.
24. Catignani GL, Bieri JG: Simultaneous determination of retinol and α-tocopherol in serum or plasma by liquid chromatography. Clin Chem 29:708, 1983.

25. Chou PP, Jaynes PK, Bailey JL: Determination of vitamin E in microsamples of serum by liquid chromatography with electrochemical detection. Clin Chem 31:880, 1985.

26. MacCrehan WA, Schonberger E: Determination of retinol α-tocopherol, and β-carotene in serum by liquid chromatography with absorbance and electrochemical detection. Clin Chem 33:1585, 1987.

27. Davis GC, Kissinger PT: Strategies for determination of serum or plasma norepinephrine by reverse-phase liquid chromatography. Anal Chem 53:156, 1981.

28. Parker NC, Levtzow CB, Wright PW, Woodard LL, Chapman JW: Uniform chromatographic conditions for quantifying urinary catecholamines, metanephrines, vanillmandelic acid, 5-hydroxyindole acetic acid, by liquid chromatography, with electrochemical detection. Clin Chem 32:1473, 1986.

29. Kodama K, Yamanaka K, Nakata T, Aoyama M: Liquid-chromatographic assay of urinary vanillmandelic acid and homovanillic acid, with clean-up by on-column injection of acetonitrile or methanol. Clin Chem 32:1944, 1986.

30. Eisenhofer G, Goldstein DS, Stull R, et al: Simultaneous liquid-chromatographic determination of 3.4-dihydroxyphenylglycol, catecholamines, and 3,4-dihydroxyphenylalanine in plasma, and their responses to inhibition of monoamine oxidase. Clin Chem 32:2030, 1986.

31. Foti A, Kimura S, DeQuattro V, Lee D: Liquid-chromatographic measurement of catecholamines and metabolites in plasma and urine. Clin Chem 33:2209, 1987.

32. Caldarella AM, Reardon GE, Canalis E: Analysis for cortisol in serum by liquid chromatography. Clin Chem 28:538, 1982.

33. Andreolini F, Borra C, Caccamo F, et al: Estriol and its conjugates in late pregnancy determined by extraction with carbopack B and liquid chromatography with fluorometric detection. Clin Chem 31:1698, 1985.

34. Turpeinen U, Koivunen E, Stenman U: Liquid-chromatographic determination of β_2-microglobulin, α_1-acid glycoprotein, and albumin in urine. Clin Chem 33:1756, 1987.

35. Biggs HG, Gentilcore LJ: Liquid-chromatographic measurement of amino acids in biological samples after formation of phenylthiohydantoin derivatives. Clin Chem 30:851, 1984.

36. Okuda T, Oie T, Nishida M: Liquid-chromatographic measurement of creatinine in serum and urine. Clin Chem 29:851, 1983.

37. Zhiri A, Houot O, Wellman-Bednawska M, Siest G: Simultaneous determination of uric acid and creatinine in plasma by reversed-phase liquid chromatography. Clin Chem 31:109, 1985.

38. Little RR, England JD, Wiedmeyer H, et al: Interlaboratory standardization of glycated hemoglobin determinations. Clin Chem 32:358, 1986.

39. Drexel H, Klocker H, Patsch JR, Breier C, Braunsteiner H: Measurement of glycated protein by a rapid and specific method for absolute quantification of lysine-bound glucose. Clin Chem 33:1656, 1987.

40. Harsing Jr LG, Nagashima H, Duncalf D, Vizi EZ, Goldiner PL: Determination of histamine concentrations in plasma by liquid chromatography/electrochemistry. Clin Chem 32:1823, 1986.

41. Johnson PM, Perkins SL, Kennedy SW: A high-speed liquid-chromatographic method for measuring urinary porphyrins. Clin Chem 34:103, 1988.

42. Buchanan DN, Theone JG: Photodiode array detection for liquid chromatographic profiling of carboxylic acids in physiological fluids: 3-hydroxy-3-methylglutaric aciduria. Clin Chem 32:169, 1986.

43. Kodama C, Ototani N, Isemura M, Aikawa J, Yosizawa Z: Liquid-chromatographic determination of urinary glycosaminoglycans for differential diagnosis of genetic mucopolysaccharidoses. Clin Chem 32:30, 1986.

17
Electrophoresis and Isoelectric Focusing

Electrophoresis

Electrophoresis describes the migration of charged particles in an electric field. The driving force (F) on the particle that causes the particle to migrate is the product of the effective charge on the particle (Q) and the field strength or applied voltage (E) expressed in volts per centimeter. Thus,

$$F = QE$$

Under conditions of constant velocity, the migrating particle will encounter resistance because of the viscosity (η) of the solvent. This frictional resistance (f) in free solution can be expressed by Stokes' law, which states that $f = 6\pi r V \eta$, where r is the ionic radius of the particle, V is the velocity with which the particle is moving expressed in centimeters (cm) per second (sec), η is the viscosity of the electrophoresis buffer, and π is a constant that equals 3.1416. The fact that the driving force on the particle (F) is balanced by the frictional resistance (f) leads to constant velocity.

In practice, Stokes' law is not strictly followed when gels are used as electrophoretic support medium, and factors such as gel density and particle size, among other factors, determine the frictional resistance (f).

The electrophoretic mobility expressed by the symbol μ is related to the velocity of the migrating particle (V), which is the distance migrated in centimeters (cm), per time in seconds (sec) under the influence of an applied electrical field (E) expressed in volts per centimeter. Thus,

$$\mu = \frac{V}{E} = \frac{\frac{\text{cm}}{\text{sec}}}{\frac{\text{volt}}{\text{cm}}} = \frac{\text{cm}^2}{\text{volt-sec}}$$

Thus the units of electrophoretic mobility μ are cm^2/volt-sec. Electrophoretic mobility can be directly related to the charge of the particle (Q) and inversely related to the frictional resistance encountered during migration, which is governed by the viscosity of the buffer medium (η) and the ionic radius or shape of the molecule (r). In other words, at constant velocity F=f, or QE=$6\pi rv\eta$. Thus,

$$\mu = \frac{V}{E} = \frac{Q}{6\pi r\eta}$$

Factors Affecting Migration

Charge: The greater the number of charges on a particle, the greater the relative distance migrated. The direction of migration will depend on the sign, with the positively charged particles moving toward the negative electrode (cathode) and the negatively charged particles moving toward the positive electrode (anode).

Time: Other conditions being equal, increasing the time of electrophoresis will increase the distance migrated by the charged particle.

Voltage: Increasing the applied voltage increases the migration distance of the charged particle.

Distance between electrodes: Increasing the distance between electrodes while keeping the applied voltage constant has the effect of spreading the voltage over a larger distance, thus reducing the effective voltage and decreasing the distance migrated by the charged particle.

Conversely, decreasing the distance between the electrodes while keeping the applied voltage constant leads to an increase in the effective voltage, thus increasing the distance migrated by the charged particle.

Buffer ionic strength: Increasing the ionic strength of the buffer leads to decreased migration of the charged particle to be separated, because the current is preferentially carried by the buffer ions. The increased ionic cloud of buffer ions around the charged particle impedes the migration of the latter. The increased conductivity of buffer ions in high ionic strength buffers leads to increased heat generation. Increased heat causes evaporation of buffer from the electrophoretic support medium. This loss of buffer from the electrophoretic support medium is replenished by the flow of buffer from the buffer compartments, thus affecting the migration of the charged particle, a phenomenon referred to as "wick flow." Increased heat generated by high ionic strength buffers leads also to an increase in the rates of diffusion of the ions.

In contrast, use of low ionic strength buffers leads to increased migration of charged particles to be separated.

In practice, although high ionic strength buffers yield sharper separations, the attendant heat generated in such buffers dictates a compromise in ionic strength. Buffer ionic strength in the range 0.025 to 0.1 is used in electrophoretic separations. Ionic strength of barbital buffers used in protein electrophoresis range from 0.025 to 0.075. Ionic strength of Tris-boric acid-EDTA buffers range from 0.03 to 0.1

The heat generated in high ionic strength buffers can lead to denaturation of proteins if a cooling system is not used. The problem with cooling devices, however, is the creation of a temperature gradient caused by the fact that some portions of the support medium may be cooled better than others. This leads to distortions in shapes of separated bands created by differences in migration rates across the length of the electrophoresis support medium.

Heat generated by high ionic strength buffers causes variation in both current and voltage, which can be minimized by use of constant current or constant voltage power supplies. The use of constant power supplies ensures that heat generated is constant throughout the electrophoretic procedure, although it still does not obviate the fact that the heat might cause denaturation of separated proteins in the absence of cooling devices to remove the heat produced.

Buffer pH: The pH of the electrophoresis buffer dictates the extent of ionization of particles to be separated. At the alkaline pH of 8.6 used for protein electrophoresis, all proteins have a net negative charge and migrate toward the anode.

Size of the molecule: For the same unit charge, a larger molecule will migrate slower than a smaller molecule.

Electroendosmosis: The electrophoretic support medium may have charged groups, such as the carboxyl groups of paper, cellulose acetate, and the sulfate groups of agaropectin in agar. In neutral or basic pH, these charged groups will be ionized and have a tendency to migrate toward the anode. However, because the negatively charged groups are fixed on the electrophoretic support, they are unable to migrate. To compensate for this, there is a movement of H^+ ions in the form of hydrated protons H_3O^+ toward the negative electrode cathode, thereby resulting in a movement of buffer toward the cathode. The effect of electroendosmosis is to speed up the movement of neutral and positively charged species toward the cathode, facilitated by buffer flow in that direction, since neutral and positively charged species swim with the current. In contrast, the migration of the negatively charged species toward the anode is impeded because of the electroendosmotic counter-current flow.

Electroendosmosis can be measured by incorporating a neutral molecule such as blue dextran with the sample and measuring the distance migrated from the point of sample application (origin) to the cathode.

Electroendosmosis is minimal with polyacrylamide and starch gels. Agar without agaropectin (purified agarose) has minimal electroendosmosis.

Miscellaneous factors: An egg-shaped molecule with increased ionic radii will migrate relatively faster than a molecule that is spherical and has decreased ionic radii.

The sample constituents should be soluble in the electrophoretic buffer to permit migration.

Adsorption of sample constituents to the electrophoretic support medium will impede migration and as such is undesirable.

Instrumentation

Typically, the apparatus consists of electrode vessels to hold buffer, a support to position the electrophoresis medium, an insulating transparent cover, and a power supply.

Figure 17–1 is a schematic representation of a typical horizontal electrophoresis apparatus.

Figure 17–1

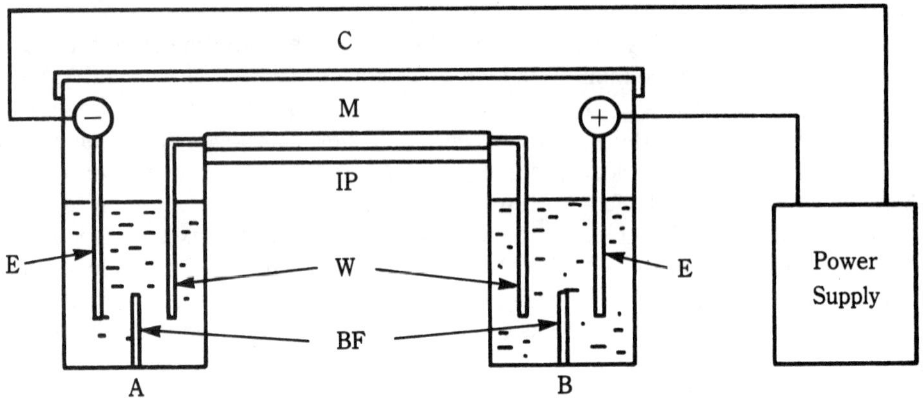

Schematic representation of a horizontal electrophoresis apparatus. A and B = buffer compartments; BF = baffle; W = wick; IP = insulating plate; M = electrophoretic support medium resting on IP; E = electrodes; C = insulating cover.

It consists of two buffer compartments. Both buffer compartments have baffle plates and contain buffer. The electrodes in the two buffer compartments are connected to a power supply. The composition of electrodes vary. Carbon rod electrodes, although low in cost, polarize readily and disintegrate with use, thus contaminating the buffer. Stainless steel or silver/silver chloride electrodes may be used, although with the former some buffers are corrosive. In general, however, platinum electrodes are preferred.

The electrophoresis support medium saturated with buffer is placed horizontally on an insulating plate. The electrophoresis medium should be free of air bubbles and excess buffer. Contact between the electrophoresis support medium and the buffer in the electrode compartments is made by wicks that dip from both ends of the support medium into the buffer. Alternatively, the ends of the support medium could dip into the buffer in the electrode compartments.

The volume of buffer used should be such that electrode products produced during electrophoresis are diluted out and do not contaminate the supporting medium during electrophoresis. Working with excess buffer allows for movement due to diffusion and convection. Electrode products are effectively prevented from reaching the electrophoretic support medium by means of baffles. Exactly how much buffer needs to be used can be calculated, if electrophoretic time and current flow in amperes and molarity of buffer are known, by use of the relationship:

$$y = \frac{I(T)}{26{,}802}$$

Where y is the gram equivalent of buffer that will be decomposed by current (I) in time (T), and where 26,802 milliampere-hour is Faraday's constant (when expressed in milliampere-hour instead of 96,487 coulombs per mole of electrons or ampere-second).

Assuming that we need a hundredfold excess of buffer, the gram equivalents of buffer required in the electrode compartments can be calculated by the equation:

$$100\ y = \frac{100\ I(T)}{26{,}802} = 0.00373\ I(T)$$

Expressed as volume of buffer of normality (N), the volume of buffer (V) needed to fill the electrode compartment can be calculated by the equation:

$$V = \frac{0.00373\ I(T)}{N}$$

After application of sample on electrophoretic support medium, the electrophoresis chamber is covered with a transparent insulating plate during electrophoresis to minimize evaporation of buffer and to insulate the apparatus.

The polarity of the electrodes is reversed after each electrophoretic run so that the anode becomes the cathode in the next run and vice versa. This allows the electrode products produced during electrophoresis to be shuttled back and forth so that they do not accumulate at any one electrode.

The power supply used should be capable of delivering a current of 50 to 100 mA or a voltage of 400 to 500 V. A surface area of 60 cm^2 on the support medium should be allowed for each 30 to 50 mA of current required.

Constant-voltage power sources are affected by the increase in current that results from the heat produced during electrophoresis. This increase in current leads to an increase in migration rate of the sample, and a decrease in resistance is further accentuated because of an increase in ion concentration caused by the evaporation of buffer from the electrophoretic support medium.

The use of a constant-current power supply solves this problem because the applied voltage is automatically decreased as the resistance is decreased, thus maintaining a constant current throughout the electrophoretic run. The migration rate is maintained relatively constant through use of a constant-current power supply, and the increase in heat during electrophoresis is minimized.

Zone Media

Paper. A variety of zone media have been utilized for electrophoresis. The use of paper for electrophoresis of serum proteins is now only of historical interest because of the inordinately long time required for the completion of separation (14 to 16 hours). Although paper has the advantage that it is mechanically strong and can be handled easily, it has several drawbacks. Albumin tends to adsorb to paper, causing trailing. The type of paper suited best for electrophoresis of serum proteins, such as Whatman No. 3MM, which is thick and soft, also has an undesirable electroendosmotic effect.

Cellulose acetate. Cellulose acetate has the advantage that its pores are homogenous and standardized from batch to batch, and that it is relatively free of the adsorption phenomena seen in paper electrophoresis. Electrophoresis time is rapid and can be completed in as little as 20 minutes. The membrane can be cleared with a mixture of methanol and glacial acetic acid and rendered transparent for densitometric scanning and subsequent storage to serve as a permanent record. A drawback of cellulose acetate is that it is brittle and must be handled with care. It is also subject to a relatively high electroendosmotic effect, although on some commercially available cellulose acetate strips, the electroendosmosis phenomenon is minimized. Because approximately 80% of the interstices of the cellulose acetate fibers are filled with air space, they swell when they are immersed in buffer. Because of their reduced buffer saturation capacity, they must be quickly transferred from the immersed buffer onto filter paper and blotted quickly without pressure and transferred to the electrophoresis chamber. The strip should be inspected to ensure that whitish spots, which indicate that air has been trapped between the pores, do not exist.

Commercially available cellulose acetate membranes vary depending on the extent of acetylation of hydroxyl groups of cellulose, the pore size, the thickness of membrane, and the prewashing step used.

Agar. Agar is made up of two constituents, agaropectin and agarose. Agaropectin has galactose sulfate and carboxylic groups, which are responsible for the significant electroendosmosis associated with agar gels. Agarose is agar without the agaropectin molecule and exhibits very little electroendosmosis. One must, however, be aware that some commercial preparations of agarose contain residual charged groups and, therefore, exhibit some degree of electroendosmosis.

Agarose. Agarose is a polymer made up of alternating units of D-galactose and 3, 6 anhydro L-galactose. It is an ideal supporting medium for the electrophoresis of serum proteins, lipoproteins, and isoenzymes. The concentration of agarose used ranges from 0.5 to 1.0 g/dL. Because of its relatively large pore size, the molecular size of a protein does not become a limiting factor. In theory, it can allow fractionation of proteins of molecular weight as high as 2,000,000 Daltons. Adsorptive effects are minimal. Being transparent, agarose gels can be readily scanned with a densitometer.[1]

Electrophoresis time is relatively rapid requiring between 30 to 90 minutes. Although sample application can be performed by a variety of techniques that range from incorporating serum in heated agar and applying the mixture to a precut well, to applying serum directly to sample well, it is best to use a template designed to eliminate sample application artifacts. The template allows delivery of a precise volume of sample (5 μL) onto each sample slot. After allowing the sample to diffuse for 5 minutes, excess sample is blotted away.

As with cellulose acetate electrophoresis serum proteins are fractionated in five fractions. In addition, the pre-albumin band is visible. Agarose provides excellent resolution of beta lipoprotein and pre-beta bands on lipoprotein electrophoresis.

High-resolution electrophoresis. The resolving power of agarose and cellulose acetate can be increased in high ionic strength buffer to yield as many as 13 serum protein fractions. In such high-resolution electrophoresis techniques, the ionic strength of the buffer used is as high as 0.075. The buffer also contains salts such as calcium lactate. The concentration of agarose used is 0.8 g/dL, the pH of the buffer remaining the same as that used for conventional serum protein electrophoresis (pH 8.6). Apparently the use of a high ionic strength buffer heightens the resolution of α, and α_2 globulin zones and albumin. The fractionation of protein fractions in the β-globulin area is facilitated by the presence of calcium lactate in the high ionic strength buffer.

Techniques exploiting endosmosis. While endosmosis in undesirable, especially in the fractionation of most lipoproteins, in some instances, such as in the visualization of lipoprotein-x, it may be desirable. Hence, agar with its component of agaropectin has been used for the electrophoretic visualization of the cathodically migrating lipoprotein-x, an abnormal lipoprotein that appears in the serum of patients with obstructive jaundice.

Agar is also used in the technique of counter immunoelectrophoresis, which exploits the endosmosis phenomena exhibited by agar. In this technique, which has been used for the separation of hepatitis β-surface antigen, the serum sample is placed between the antibody (placed near the anodal end) and the cathode. During electrophoresis, the antigen in the sample moves toward the anode, while the antibody migrates readily toward the cathode, assisted by the endosmosis flow of buffer in the same direction. As the antigen and antibody meet, an immunoprecipitin line develops indicating that the sample contains the hepatitis β-surface antigen.

Polyacrylamide. Paper, cellulose acetate, and agarose allow fractionation of molecules based on differences in charge. Media such as polyacrylamide have molecular sieving properties as well, so that separation is determined both by molecular size and charge. Polyacrylamide gel is produced by polymerizing the monomer acrylamide molecules. The resulting polyacrylamide chains are cross-linked by inclusion with acrylamide ($CH_2{=}CHCONH_2$) a bifunctional co-monomer such as *N,N'*-methylene-bis-acrylamide ($CH_2(NHCOCH{=}CH_2)_2$). The polymerization process can be effected by light (photochemically) in presence of riboflavin, or chemically with the aid of a free-radical catalyst-accelerator. Ammonium persulfate is widely used as a catalyst. Either *N,N,N',N'*-tetramethylene-diamine (Temed) or β-dimethylaminopropionitrile (DMAP) is used to start and accelerate the polymerization reaction.

The pore size of the polyacrylamide gel is determined by the concentration of the cross-linking agent (*N,N'*-methylene-bis-acrylamide) used, whereas the length of the polymer chain is dictated by the concentration of the monomer acrylamide used in the polymerization reaction.

The composition of polyacrylamide gels is widely described by the nomenclature introduced by Hjerten. The total concentration of monomer which is the sum of acrylamide and *N,N'*-methylene-bis-acrylamide in g/dL is denoted by the letter "T." The percentage by weight of total monomer "T," which can be related to the presence of the cross-linking monomer *N,N'*-methylene-bis-acrylamide, is expressed as "C."

The use of polyacrylamide gels with a T value of 3% to 30% allows fractionation of molecules with a wide range of molecular weights. Gels with a T value of 3% can be used to separate molecules with a molecular weight in excess of a million

Daltons. The 30% gel, however, is suitable for separation of molecules with a molecular weight of approximately 10,000 Daltons. As such, the total monomer concentration (T) and pore size are inversely and linearly related. It must be noted that acrylamide and β-dimethylaminopropionitrile (DMAP) are very toxic. Inhalation of acrylamide dust or contact of both of these compounds with skin must, therefore, be avoided. The polymerized gel, however, is relatively nontoxic.

Polyacrylamide gels have been used in both horizontal and vertical types of electrophoresis.

Disc electrophoresis. The fractionation of serum proteins to yield as many as 20 discrete fractions can be accomplished with discontinuous polyacrylamide gels of differing pore size and discontinuous buffer gradients. The procedure is called disc or discontinuous electrophoresis. The term disc indicates not only the discontinuities in the gel and buffer used but refers also to the discoid shape of the protein bands resulting from electrophoresis.

In the disc electrophoresis procedure, the gels (the separating gel, the stacking gel, and the loading gel) are sequentially cast in cylindrical glass tubes and polymerized. During the polymerization step, the base of the glass tube is held by a rubber cap and sits vertically in a rack. Gel components of the running or separating gel, which has a smaller pore size and a higher pH than either the loading or stacking gel, are introduced into the glass tubes, layered with distilled water to prevent inhibition of polymerization by oxygen, and polymerized under bright fluorescent light for 30 minutes. Subsequently, the distilled water layer is carefully removed with an adsorbent paper, and the gel components of the stacking or spacer gel, which has a larger pore size than the separating gel, are added, layered with distilled water, and polymerized for another 20 minutes. At the end of this period, the serum sample can be applied, mixed with a loading gel, which has the same pH as the spacer gel.

The glass tube containing the three separate gels is inserted vertically in an electrophoresis chamber containing an upper and lower buffer reservoir. The pH of the Tris-glycine buffer in the upper reservoir is 8.3, which is approximately 2 pH units higher than the pH of the spacer gel, which contains Tris-Hcl. The pH of the Tris-Hcl buffer in the lower reservoir at the start of the electrophoresis procedure is 8.9. The upper reservoir holds the negative electrode (cathode) while the lower reservoir contains the positive electrode (anode). As electrophoresis begins, the protein components in the sample move readily through the large pore size spacer gel. Glycine molecules at the pH in the upper reservoir buffer exist mainly as the zwitterion, with only 5% existing as the glycinate anion. As the glycinate anion slowly migrates into the spacer gel, its concentration is further reduced by the low pH of the spacer gel, which protonates some of the glycinate anions to yield zwitterions. This results in chloride ions moving faster than the glycinate ions, and the protein fractions become stacked in between. The discontinuity in buffer pH in the spacer gel and the upper reservoir results in stacking the protein fractions as they enter the separating or running gel as an extremely narrow band. The separating gel has a smaller pore size than the spacer gel, and separation takes place not only by charge but also by molecular size, with molecules larger than the pore size of the gel being impeded from migrating further. Since the pH of the separating gel is 8.9, which is closer to the pKa_2 of glycine, the resulting greater proportion of glycinate anions moves faster than the protein anions and forms a sharp boundary with the chloride anions. With the increase in concentration glycinate anions, the pH of the separating gel increases to pH 9.5 as the Tris-Hcl buffer is replaced by the Tris-glycine buffer system. The increased net negative charge on the proteins causes them to migrate faster, and they are limited only by their molecular size. The sharp bands produced in disc electrophoresis is due to the discontinuity of gels and gel and buffer pH.

Advantages of polyacrylamide gels. Polyacrylamide gels have several advantages. They can be produced uniformly in a variety of pore sizes. Convection currents and diffusion effects are minimal, thus providing sharp zones of separated components. Electroendosmosis and adsorption effects are also minimal. Polyacrylamide gel electrophoresis (PAGE) can be performed with a wide range of buffers. The gels can be readily cast either in cylindrical tubes or slabs. Because polyacrylamide gels themselves do not absorb ultraviolet light in the region of 270 to 280 nm, the concentration of separated protein bands can be determined by measuring their absorbance at 280 nm. By incorporating a tracking dye such as bromophenol blue with the sample, the migrating proteins can be visualized during disc electrophoresis. The molecular sieving effect that permits the separation of constituents by both their size and charge characteristics can be readily achieved with polyacrylamide gels whose pore size, in contrast to that of agarose, can be easily controlled by appropriate variations in the concentration of acrylamide and the cross-linking bis-acrylamide.

SDS—polyacrylamide gel electrophoresis. By incorporating sodium dodecyl sulfate, a detergent, with the sample, proteins can be dissociated into their subunit polypeptide chains. The dissociated polypeptide chains on polyacrylamide gel electrophoresis are separated according to their molecular weights. The rationale for the use of SDS is that it, by its high level of binding and constant amount bound for most proteins (1.4 g SDS are bound per gram of protein), neutralizes the charge on the protein, leaving behind a constant negative charge per unit mass. The negative charge is caused by the sulfate in the SDS molecule. With optimum SDS binding, the resulting protein-SDS complexes behave as rods in solution. Molecular weights of SDS-separated protein subunit chains can be determined by plotting the logarithms of the molecular weights of known polypeptide chains versus their electrophoretic mobility. From this linear plot, the electrophoretic mobility of an unknown SDS-separated protein chain can be used to determine its molecular weight. Treatment of sample with SDS is usually performed in presence of a sulfahydryl reagent such as 2-mercaptoethanol to prevent reformation of disulfide bonds. Indeed quantitative dissociation into subunit polypeptide chains is possible if sufficient quantity of SDS and 2-mercaptoethanol is added to the sample and the mixture is heated to 100°C. SDS binding may not always be optimal. The amount bound may be reduced because of the restricted conformation of the protein resulting from the inability to reduce disulfide bonds. Thus, bovine serum albumin binds only 0.9 g SDS per gram of protein. Glycoproteins exhibit decreased SDS binding because the protein portion and not the carbohydrate portion of the molecule binds SDS. The amino acid composition of a protein also determines the extent of SDS binding, with highly charged polylysyl glutamic acid chains binding as little as 0.4 g SDS per gram of polypeptide. Because of the variability in SDS binding to proteins, the molecular weight determined by SDS-polyacrylamide gel electrophoresis for most proteins can come within 10% of the true molecular weight. SDS and 2-mercaptoethanol treatment is useful not only for the dissociation of enzyme subunits, but also for the dissociation of antigen-antibody complexes. Subsequent polyacrylamide gel electrophoresis permits recovery of dissociated antigen. SDS-polyacrylamide gel electrophoresis has been used in two-dimensional electrophoresis. Proteins are first separated according to their isoelectric points (isoelectric focusing) in the first dimension, and according to their molecular weights by SDS-polyacrylamide gel electrophoresis, in the second dimension. The resolving power of two-dimensional electrophoresis is tremendous, because it has an ability to separate more than one thousand proteins on a single gel. Identification of these numerous proteins requires computerized scanning and mapping techniques.[2]

Blot techniques. Blot techniques that involve electrophoretic separation followed by transfer of separated species to sheets such as nitrocellulose are currently widely used in molecular biology and immunology laboratories.[3]

The "Southern blot technique" involves: separation by agarose gel electrophoresis of DNA fragments obtained using restriction endonuclease enzymes; subsequent transfer of the electrophoretically separated fragments onto nitrocellulose sheets; and hybridization to a complementary strand of radiolabelled (32_p) RNA. Radioactivity can be detected by exposing the hybridized DNA fragment to X-ray paper (autoradiography) and visualizing the darkened spots.[4]

Separation of RNA by agarose gel electrophoresis and transfer to nitrocellulose sheets is the basis of the "Northern blot" procedure. Sensitive blot techniques such as the "Western blot" technique have been used for the detection of AIDS anti-HIV antibodies.[5] In this procedure, disrupted and inactivated AIDS virus (HIV) is separated into its component proteins by SDS-polyacrylamide gel electrophoresis and is electrophoretically transferred to nitrocellulose strips. The strips are incubated with diluted serum sample. If HIV antibodies are present in serum, they will bind to the electrophoretically separated AIDS viral protein fragments. After washing to remove the unbound material, an enzyme-linked antibody to human I_gG is added. Unreacted enzyme-linked antibody is removed by washing. Substrate is added, and the appearance of colored bands indicates the presence of HIV antibody in the serum sample.

Blotting techniques have the advantage that transfer to an appropriate matrix such as nitrocellulose, or diazobenzyloxymethyl paper (DBM), not only preserves the identical electrophoretic pattern obtained in the gel from which transfer was effected, but it also permits use of a variety of techniques to visualize the patterns. Once transferred to the matrix, diffusion of bands is no longer a problem. Blotting techniques can also be used in conjunction with high resolution electrophoresis techniques such as two-dimensional electrophoresis and isoelectric focusing. Because of the emerging importance of blotting techniques, it is appropriate to highlight a few details.

Electroblotting. Currently the electrophoretic transfer of proteins from gels to a matrix (electroblotting) appears, because of its efficiency, to be the method of choice.[6,7] A wide range of electroblotting apparatus is commercially available. Briefly, electroblotting transfer is accomplished by sandwiching the gel containing the separated proteins and an appropriate matrix such as nitrocellulose between filter paper and sponge pads, which are inserted between the electrodes. The electrodes are then inserted into a reservoir containing buffer, and electrophoretic transfer is effected by applying an electrical field that should ideally be homogeneous over the entire area of the gel. Usually transfer time is approximately 5 hours. If acid buffers are used, the nitrocellulose matrix should face the cathode. When alkaline buffers (Tris/glycine, pH 8) are used as with SDS gels, the matrix should face the anode.

The binding of nitrocellulose to protein is apparently caused by hydrophobic interactions. Although nitrocellulose continues to be a widely used matrix for electroblotting, the use of DBM paper provides the advantage of covalent protein binding. The disadvantage, however, is that prior to use, there is a pretreatment diazotization step to convert aminobenzyloxymethyl-paper to diazobenzyloxymethyl (DBM) paper. At the completion of electroblotting, nonspecific binding sites on the protein are blocked by a suitable blocking agent such as Tween 20, a detergent. The blocking step minimizes the interaction with the matrix of specific reagents such as antibodies or ligands that may be used to detect the transferred protein, and thus limits the background interference.

Visualization. Visualization of transferred constituents on the matrix can be achieved by use of a wide variety of labels, of which nonisotopic enzyme labels, for obvious reasons, are the method of choice.

While direct visualization of the transferred constituent by use of a specific enzyme label is the most convenient approach, in practice a second or even third label is generally used. In this indirect immunolabelling technique, the antigen on the matrix is first reacted with a specific label or antibody. For detection, a second labelled antibody is added that combines with the primary antibody bound to the antigen on the matrix. Alternatively, the second antibody added is not labelled, and a third peroxidase-labelled antibody or a preformed peroxidase-antiperoxidase complex (PAP) as third antibody is added for detection. The PAP technique has less than 0.1 nanogram sensitivity and is widely used. Alkaline phosphatase-labelled antibody with indoxyl phosphate as substrate, and immuno gold-silver stains, the latter having picogram sensitivity, are some of the other alternative labels for use with blotting techniques.

The applications of blotting techniques can range from detection of hormones, receptors, enzymes, viral antigens, allergens, autoantibodies, DNA-binding proteins, apolipoproteins, glycoproteins to oncogenes. The use of biotin conjugated lectins allows visualization of glycoproteins on the matrix through the use of avidinperoxidase label, since avidin binds to biotin. The expanding literature on blotting techniques is testimony to their versatility.

Quantitation of Separated Electrophoretic Fractions

The amount of dye bound to a separated protein fraction can be measured either by densitometry or elution of the dye with appropriate solvent and subsequent measurement of the eluted dye by spectrophotometry.

Densitometry. A densitometer is essentially a double-beam filter photometer or spectrophotometer that scans the electrophoretic strip as it moves past the optical system. The sample beam scans the dye intensity of the separated fraction, whereas the reference beam scans the area of the strip that is devoid of separated material and thus computes the dye background intensity. The difference between the reference and the sample detector signals are presented as a recorder tracing. The area under each densitometer peak can be computed with either a mechanical integrator, or as is currently available, densitometers by use of an electronic integrator. Distinction between peaks is made by sensing the change in current at the end of each peak. The densitometer electrical signal can be converted by an analog to digital converter to print out percentage of area under each peak relative to the sum of the areas, or total of all peaks. Entering the total protein value into the densitometer keyboard allows printout of the absolute concentration of each of the separated fractions.

A densitometer should, ideally, compensate for the electrophoretic strip background (automatic zeroing) to insure that peaks representing very high concentration are still maintained on scale by use of an automatic gain control. A densitometer should also be able to scan electrophoretic support materials varying in length between 25 and 150 mm, at variable wavelengths in the range 350 to 700 nm, with preferably an additional wavelength capability at 254 nm, and with variable slits ranging from 0.1 mm wide to 10 mm high. Densitometers should also have fluorescence capability to measure fluorescence of NADH and NADPH-coupled enzyme reactions used for the quantitation of electrophoretically separated isoenzyme bands.

The introduction of automated electrophoresis instruments has simplified the time-consuming manual steps and provided extra options, such as the ability to scan electrophoretograms at a variety of scanning speeds, to store and recall many electrophoretograms, and to troubleshoot instruments with microprocessor-controlled algorithms.

Quality Control

Each step in the electrophoretic procedure from sample application to staining and densitometric scanning can give rise to errors and, therefore, must be controlled.

Overloading of sample can cause adsorption of protein to electrophoretic zone media and lead to distortion of separated bands and trailing. Furthermore, the intensity of dye absorption associated with such distorted bands limits the ability of most densitometers to accurately compute their concentrations. As such, the repeatability of sample applicator should be assessed by applying eight replicate applications of a single serum sample and comparing the absolute concentrations of each of the electrophoretically separated fractions.

Sample application artifacts can arise if the applicator is pressed too hard on the electrophoretic support, which traps protein that may become adsorbed at the point of application. Artifacts can also occur by the use of dirty applicators, which may introduce denatured protein onto the support medium. A broad gamma globulin band with a notch indicates an apparent biclonal origin and may be an artifact of sample application. It can be seen by examining the stained membrane for a notch or cleft in the middle of the gamma globulin band.

Drugs bound to albumin can distort the albumin band and may even split the albumin peak thus producing a spurious bisalbuminemia. Bilirubin binding to albumin can cause artifactual fluorescence and thus might interfere in the fluorometric scanning of isoenzyme bands such as those of creatine kinase. Reaction of albumin with acidic or alkaline buffers can also give rise to additional spurious bands.

Unequality of binding of stains to separated protein fractions is a problem. Ideally, all separated protein fractions should bind dye equally on the basis of weight, but in practice this is rarely the case. Thus the protein stain Ponceau S has a greater number of binding sites for albumin than for other serum proteins. Because albumin also binds to drugs, toxins, and bilirubin, the albumin band may be underestimated in drug overdose or in renal failure when toxins saturate albumin binding sites.

Staining conditions such as temperature and time of staining should be carefully controlled. The fixation of proteins should be complete prior to staining. Incomplete fixing can underestimate protein fractions due to solubility and loss of protein fractions during staining manipulations. The strip should preferably be agitated during staining to insure complete staining. Care should be taken to prevent loss of stain from protein fractions during washing steps intended to eliminate the background dye.

Densitometer slit width is a variable. The use of slit widths narrower than the width of the narrowest zone will result in overestimation of such fractions. However, protein bands whose widths are wider than the scanning slit width will be underestimated.

Incompletely cleared cellulose acetate membrane will interfere with densitometric scanning because of the membrane's opacity. For this reason, cellulose acetate membranes should be completely cleared with an organic solvent prior to densitometric scanning.

The densitometer performance should be checked with standards that permit the pen to return to baseline between peaks. Photographically produced optical density wedges accurate to within 10% can be used to calibrate densitometers. It is important to scan a single stained membrane at each instrument setting, changing one setting at a time, such as scanning speed, slit width, or filter wavelength. Differences in scanning speeds can contribute to imprecision. The positioning of the membrane with respect to the light path can also affect reproducibility.

Because variations exist between membranes of the same lot, a good quality control scheme to insure that the procedure is in control is to apply a single sample eight times to three different membranes of the same lot number (a total of 24 samples). It is best to repeat the above each day for five consecutive days, preferably by different technologists. Finally, a new lot of membranes and reagents should be evaluated by the above scheme to insure that the electrophoresis procedure is in control.

Isoelectric Focusing

Separation of the constituents in isoelectric focusing is based on their differences in isoelectric point. A pH gradient is established by using an heterogeneous mixture of aliphatic polyamino polycarboxylic acids, which are referred to as ampholytes. The ampholytes are mixed with sucrose to provide a dense, convection-free medium, or they may be incorporated with a medium such as cellulose acetate, polyacrylamide, or agarose. A pH gradient is established as the carrier ampholytes are driven in an electric field based on the sign of their charge toward the appropriate electrode. The electrode at the anode is an acid such as phosphoric acid, whereas at the cathodal end, the electrolyte used is a base such as ethanolamine. The pH gradient results because the ampholytes can migrate only up to a point where the pH equals the isoelectric point of a particular carrier ampholyte. At the isoelectric point, the charge on the migrating particle is zero and no further migration is possible. By using an appropriate mixture of carrier ampholytes, either a broad pH gradient such as pH 3 to 10, or a narrow pH gradient with a range of 2 pH units (eg, pH 3 to 5, 4 to 6, etc) can be created.

If a sample such as a mixture of isoenzymes is applied, the isoenzymes will migrate in the pH gradient created by the carrier ampholytes until they reach a pH corresponding to each of their respective isoelectric points. At this point, the isoenzyme cannot migrate further, and it becomes focused. The tendency of the constituent to diffuse from either of the focused zones is prevented by the acquisition of charge that will drive it back again to the focused zone. Thus if the focused protein tends to diffuse toward the cathode, it will acquire a negative charge and will be driven back toward the original zone. Likewise, if the focused protein tends to diffuse out toward the anode, it will acquire a positive charge, and thus will be quickly shuttled toward the original zone. Isoelectric focusing has such resolving power that it can separate constituents whose isoelectric points (pI) differ by 0.02 pH units.

Isoelectric focusing is usually carried out with a constant-power power supply, which is preferred over the constant-voltage condition. In this condition, voltage has to be adjusted as the current drops because of the lower conductivity of ampholytes as they attain their isoelectric pH, and because of the accumulation of water on the surface of the medium during electrophoresis. The concentrations of ampholytes used in isoelectric focusing necessitate cooling of the apparatus because of the high voltages (in the vicinity of 1100 to 2000 V) that occur during isoelectric focusing.

Carrier ampholytes chosen for isoelectric focusing are characterized by their possessing both good buffer capacities and conductivities and negligible electrophoretic mobility in relation to the sample constituents separated. Because their molecular weights are generally between 300 to 600, they can easily be removed from the separated protein or other large molecule weight constituent by dialysis. Proteins separated by isoelectric focusing can be quantitated by the measurement of their UV absorbance at 280 nm since the carrier ampholytes do not absorb in that UV wavelength. The ampholytes chosen are not chemically reactive with the constituents in the sample.

Types

Isoelectric focusing can be performed in vertical mounted columns or in horizontal gel media such as cellulose, polyacrylamide, or agarose. While the column technique is time-consuming, isoelectric focusing in gels is relatively rapid, taking between 1 and 3 hours.

It is important, however, that gels used in isoelectric focusing have minimal electroendosmotic flow to minimize the accumulation of water on the gel surface during electrophoresis. Gels used should be optically clear and amenable to quick staining and destaining procedures. The gels should be capable of being pressed and dried to be maintained for record purposes. They should have large pores so that they do not impede the migration of large molecules by molecular sieving.

Although polyacrylamide, cellulose acetate, and agarose in general satisfy the above criteria, some investigators have cited preference of one gel over another. Thus nontoxicity of agarose over polyacrylamide, and the ease of preparation of the former, has been cited in the preference for agarose.[1] The need for multiple chemicals, lamps for photopolymerization, and problem of gel formation in certain pH ranges have been cited as limitations for polyacrylamide when compared to agarose. The agarose gel, it has been pointed out, has much larger pore size relative to polyacrylamide and can be used to fractionate constituents with molecular weights as high as 2,000,000. In contrast to agarose, catalysts and residues of gel-forming chemicals associated with polyacrylamide may be inhibiting to some enzymes. Finally, agarose gels are amenable to quick staining and destaining. Compared to polyacrylamide, the procedure for preparation of dried gels is rapid and relatively simple.

Limitations

Although diffusion of focused protein is prevented during electrophoresis, once the electrical field is switched off, diffusion can occur and obscure the focused zone. This necessitates prompt staining and fixing of the focused constituents.

Precipitation of proteins may occur due to their low solubility in the region of their isoelectric pH and thus interfere with the proper formation of the pH gradient and the resultant separation. This can be obviated by increasing either the concentration of the carrier ampholytes or by adding urea, both of which serve to increase protein solubility.

Applications

Isoelectric focusing has been used to separate isoenzymes (eg, alkaline phosphatase),[8] hemoglobin variants, and variants of specific proteins, such as α1 antitrypsin. It can also be used to establish conditions for ion-exchange chromatography.[9] If the isoelectric point is known, the constituent that needs to be separated can be applied to a cation exchange column at a pH just below the isoelectric point of the protein. Here it will be positively charged and eluted by increasing pH of buffer to the pH that corresponds to the isoelectric point of the protein. Conversely, the protein can be applied to an anion exchange column at a pH just above the isoelectric point of the protein. Here it will be negatively charged and eluted by reducing pH of buffer to a pH equal to the isoelectric point of the protein. At the isoelectric point, a protein will not be charged and therefore will not be retained on an ion exchange column and can thus be easily separated.

References

1. Radola BJ (ed): Electrophoresis 70, Advanced Methods and Clinical Applications. New York, Walter de Gruyter, 1980.
2. Hochstrasser D, Augsburger V, Pun T, et al: "High-resolution" mini-two-dimensional gel electrophoresis automatically run and stained in < 6h with small, ready-to-use slab gels. Clin Chem 34:166, 1988.
3. Victor C, Tsang W, Peralta JM, Simmons AR: Enzyme-linked immunoelectrotransfer blot techniques (EITB) for studying the specificities of antigens and antibodies separated by gel electrophoresis. Methods Enzymol 92:377, 1983.
4. Southern EM: Detection of specific sequences among DNA fragments separated by gel electrophoresis. J Mol Biol 98:503, 1975.
5. Burnette WN: "Western blotting": Electrophoretic transfer of proteins from sodium dodecylsulfate-polyacrylamide gels to unmodified nitrocellulose and radiographic detection with antibody and radioiodinated protein A. Anal Biochem 112:195, 1981.
6. Beisiegel U: Protein blotting. Electrophoresis 7:1, 1986.
7. Dunn MJ (ed): Electrophoresis, 86. VCH Publishers, Deerfield Beach, Florida, 1986.
8. Griffiths J, Black J: Separation and identification of alkaline phosphatase isoenzymes and isoforms in serum of healthy persons by isoelectric focusing. Clin Chem 33:2171, 1987.
9. Lin YT, Li SC: Some experiments in the fractionation of proteins by isoelectric focusing. Ann NY Acad Sci 209:187, 1973.

18
Immunoanalytical Techniques

The ability of an analyte (antigen) to react with an antibody directed to it is the basis of immunoanalytical measurements. Essentially, an antigen-antibody complex is the result of the combination of a bivalent antibody molecule, such as IgG, with a multivalent antigen molecule toward which the antibody is directed. The primary phase of the reaction of antigen and antibody is the formation of an immune complex that scatters light. This phase is independent of the electrolyte concentration. The secondary phase of the reaction is the formation of an immune precipitate as antigen-antibody complexes form a cross-linked lattice. The full course of the precipitin reaction is traced by adding increasing concentrations of antigen to a constant amount of antibody. The diameter of the precipitin ring that results for each increment of antigen addition is quantitated, if the reaction is performed in a flat agar-impregnated plate. If the reaction is performed in a test tube, the nitrogen content of the precipitate is determined. A plot of the precipitin ring diameter or the nitrogen content of the precipitate vs the concentration of antigen added is illustrated in Figure 18–1.

Initially there is an excess of antibody. As antigen is added to the antibody, the amount of immunoprecipitate formed increases until all the antibody is in combination with the added antigen, and the maximum amount of immunoprecipitate is formed. At the equivalence point, antigen and antibody molecules form a structured lattice network. If the reaction is performed in a test tube, and the immune precipitate is centrifuged, there will be no antigen or antibody remaining in the supernatant. Further incremental additions of antigen beyond the equivalence point result in the break-up of the structured lattice and the dissolution of the immune complex. For this reason, immunoanalytical techniques dependent on the properties of an immune complex, such as light scattering or the formation of measurable precipitin ring, are performed under conditions of antibody excess.

Figure 18–1

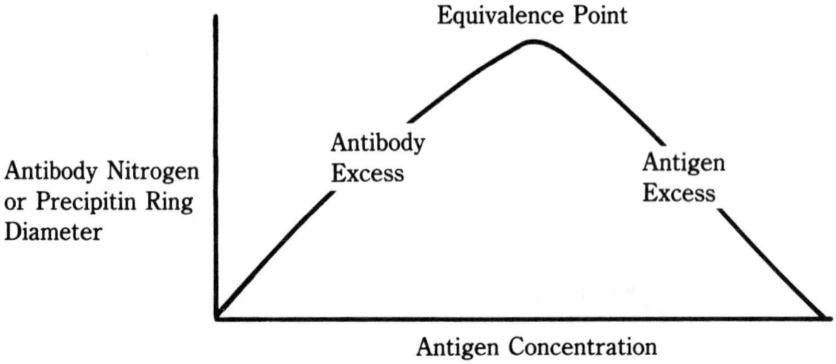

Precipitin reaction. Increasing amounts of antigen are added to fixed antibody concentrations in separate experiments. The amount of nitrogen in the immune precipitate or the diameter of the precipitin ring is measured and plotted against the antigen concentration.

The primary phase of the antigen-antibody reaction can be exploited using either isotopic (radioimmunoassay) or nonisotopic labels (enzyme or fluoroimmunoassays).

The forces that hold the antigen-antibody complex are weak forces such as interaction with electron clouds (Vanderwaals forces), hydrogen bonding, hydrophobic interactions, and ionic or coulombic forces. An idea of how weak these forces are can be appreciated when the strength of a Vanderwaals force bonding is 0.5 kcal, and the hydrogen bond strength is 2 to 5 kcal compared to a covalent bond strength of 40 to 140 kcal.

The sensitivity of an immunoassay is dependent on the strength of antigen binding through its binding site (antigenic determinant or epitope) to a single antibody binding site. This property characterizes antigen affinity. Because an antibody has more than one binding site (IgG has 2, whereas IgM has 10), the strength of binding of antigen to the number of antibody binding sites is a measure of the avidity of antibody. Polyclonal antibodies have increased avidity in contrast to monoclonal antibodies, which are directed toward a single specific epitope on a multivalent antigen molecule. Because of the specific interaction of monoclonal antibody with a single epitome on the antigen molecule, such complexes cannot cross-link to form larger complexes that precipitate. As such, they are limited in procedures relying on immunoprecipitation.

The strength of antigen-antibody binding can be decreased by increasing ionic strength. The ability of ions to interfere with antigen-antibody binding is related to their ionic radius and degree of hydration. In general, the greater the ionic radii and the lower the degree of hydration of ion, the greater is the interference with antigen-antibody binding. Cations interfere with the binding of cation binding group on the antigen molecule with anion binding group on the antibody. Anions interfere with the binding of anion binding group on the antigen molecule with the cation binding group on the antibody. Antigen-antibody complex precipitation can be enhanced by inclusion of polymers that decrease the solubility of such complexes and hasten their precipitation. This is the basis of the use of water-soluble linear high molecular weight polymers such as dextran, polyethylene glycol, or polyvinylpyrrolidone to speed up antigen-antibody reactions.

Techniques That Exploit Properties of Antigen-Antibody Complexes

Nephelometry

Antigen-antibody complexes formed in the primary phase of the precipitin reaction scatter light. Measurement of light scatter of the immune complex is the basis of nephelometric measurements. Earlier applications of nephelometry to immunoglobulin quantitation were based on the standard practice of measuring light scattering at right angles to the incident beam (90° angle). This was performed in a nephelometer or a fluorometer with a similar configuration of the detector to the incident beam, without using a secondary filter or a secondary monochromator. The drawback of measuring light scatter at a 90° angle is that discrimination of background light scatter, due to the antigen-antibody complex, is difficult. Despite this and other drawbacks such as increased cost, unreliability in antigen excess, and difficulty in keeping the system clean, automated immunoprecipitin techniques based on the measurement of 90° angle light scatter have been used routinely in several laboratories for the measurement of immunoglobulins[1,2] and have been adapted to centrifugal analyzers.[3,4]

The utilization of laser beams as a light source in nephelometry has permitted measurement of light scattering at angles of less than 90°.[5] In one such system a helium/neon laser emitting photons at 632.8 nm is used as a light source, and forward light scatter produced by the antigen-antibody complex is detected at an angle of 31° by a photomultiplier tube.[5] This system has the advantage of eliminating background light scatter associated with dust and other extraneous particles, and electronically subtracting scatter caused by the sample, antibody, and reagent buffer from the antigen-antibody scatter. The use of a polymeric buffer maintains antigen-antibody complexes in an insoluble form, permitting measurement of relative light scatter percentage that is directly proportional to the concentration of antigen in the analyzed sample. Typically, antibody and sample are incubated for one hour before taking light scattering measurements. This is the timed end-point or equilibrium approach. Light scattering may be measured, however, at intervals between 15 and 60 minutes.[5] Precise timing to within 15 seconds is required when taking measurements after only 15 minutes incubation, as there is a likelihood that the reaction may not have reached a steady-state. Taking measurements after 60 minutes incubation without shaking the assay mixture could result in reduced light scatter due to the sedimentation of the antigen-antibody complex.[5]

Nephelometers that permit measurement of the maximum rate of change of forward light scatter are also available. In a system based on this principle of measurement, the light source, a tungsten/iodine lamp, and the detector, a photomultiplier tube, is positioned at a 70° forward angle. A wavelength range of 400 to 550 nm is used to measure light scatter. The reaction is triggered by the introduction of the appropriate antiserum, containing a dye, to a continually stirred mixture of diluted sample and polymer enhancer in a disposable cell. A photodiode senses the trigger dye. While the absorption of the triggered dye is determined at a longer wavelength to signal the initiation of the reaction, the maximum rate of change of forward light scatter of the antigen-antibody complex produced within 60 seconds of the reaction initiation is used by microprocessor to determine the antigen concentration.[6] Specimens in the antigen excess zone are identified by the instrument whenever the rate signal drops to a predetermined level and are checked by an additional dispensing of the calibrator solution. The measurement of the maximal rate of change of forward light scatter eliminates the need for including a sample or reagent blank and allows results to be computed based on a single-point calibration. Because the time allowed to mix the sample and antibody in the disposable cell is as short as 20 seconds, timing becomes critical.

Precision, in general, is reported to be comparable for the three types of nephelometers discussed above.[7] Inadequacy of mixing, samples in antigen excess, and the lack of precision of timing end point reactions that have not reached steady-state are variables that affect nephelometric measurements. The nonspecific absorption of immunoglobulins to some of the filters used to clarify turbid sera will also affect nephelometric measurements. This effect is reported to be minimal with a polycarbonate type of filter.[8]

Immunodiffusion

Immunodiffusion is one of the many immunoanalytical techniques that utilizes the secondary phase of the antigen-antibody reaction in which an immune precipitate is formed as antigen-antibody complexes cross-link and precipitate.

Immunodiffusion can be a passive process in which one or both reactants diffuse passively toward each other, or it can be an active diffusion under the influence of an electrical field applied through the gel matrix. Single and double gel diffusion tests are examples of passive diffusion.

Single diffusion. Single diffusion of antigen or antibody through an agarose or agar gel matrix results in the formation of an immune precipitin ring. If the diffusion is performed in a test tube where the gel matrix contains the antibody, and the antigen layered over the gel diffuses in one direction, the procedure is referred to as linear or one-dimensional, single diffusion. This procedure is the basis of the classical Oudin technique.[9] If single diffusion is performed in an antibody-coated gel matrix in a plate, the antigen aliquoted into a well in the matrix is allowed to diffuse radially and form a precipitin ring. This radial immunodiffusion procedure (RID) is an example of a two-dimensional, single diffusion. (Figure 18–2)

The immunoglobulins IgG, IgA, and IgM and a variety of clinically useful immunochemical analytes can be routinely quantitated by RID. The diameters of the precipitin ring can be measured while the ring is still growing[10] or when the reaction has been allowed to reach equilibrium.[11] The former fixed time approach is usually

Figure 18–2

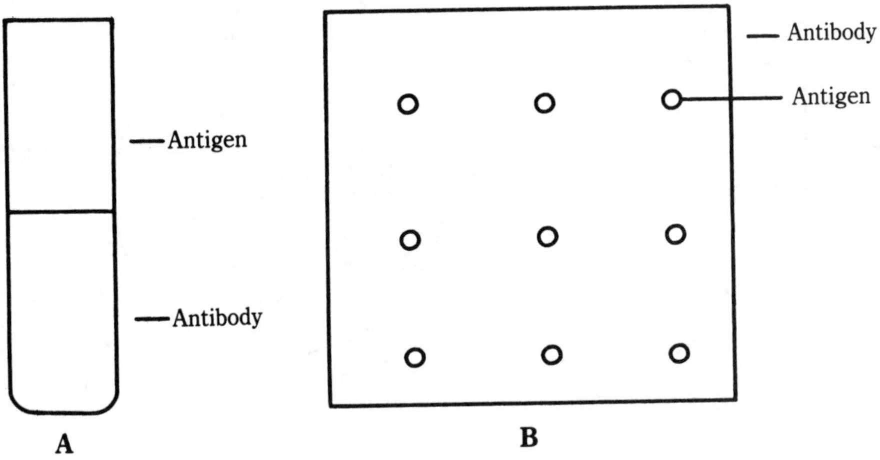

Single diffusion. A: One dimension in which antigen is layered over antibody in a test tube. B: Two dimensions in which Ag is applied to a well created in an antibody-coated gel, the classic radial immunodiffusion technique. In single diffusion, antigen and antibody are in intimate contact with each other.

performed 24 hours after commencing diffusion, whereas the latter technique can take 48 to 72 hours. A calibration curve is prepared by plotting the square of the diameters of the precipitin ring on a linear scale versus the concentrations of the standards on a logarithmic scale using linear logarithmic graph paper. The square of the diameter of unknown specimens can be read off such a curve to obtain their corresponding concentrations. The diameters of precipitin rings are usually measured in duplicate (with readings taken at two axes at right angles to one another) with the aid of a magnifier calibrated eye piece, and the two values are averaged.

The RID technique is reported to have a sensitivity ranging from 0.3 mg/dL[11] to 2 mg/dL[12] and a coefficient of variation between 5% to 7%.[12] Attention to accurate pipetting and proper filling of antigen wells, the exactness of timing, and the accuracy in measurement of ring diameters all contribute to improved precision. Because of their long-standing use, RID procedures are used to evaluate newer immunochemical techniques.

Because diffusion is related to molecular size, however, values obtained by RID may not always correlate to values obtained by techniques independent of the molecular size of the antigen. Thus, samples containing low molecular weight monomeric IgM have been reported to give apparently higher values than those obtained by laser nephelometry. In contrast, samples containing polymeric IgA have been reported to give apparently lower values by RID compared with those obtained by laser nephelometry.[13]

In one study significantly higher values were obtained for IgG, IgA, IgM and C_3C, and significantly lower values for C_4 by laser nephelometry compared to RID.

Figure 18–3

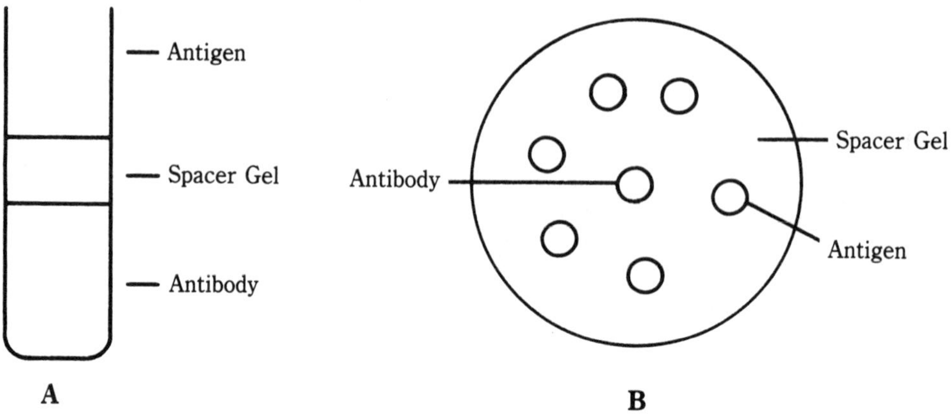

Double diffusion in which a spacer gel separates the antigen and antibody. A: One dimension. Antigen layered over a spacer gel underneath which is the antibody. B: Two dimensions. Antigen and antibody separated by spacer gel in a plate or petri dish. The classic ouchterlony technique where troughs are created in a gel matrix for application of antigen and antibody is used.

Part of the bias was ascribed to the equation programmed in the data processor for calculating standard curves in nephelometry.[14] Radial immunodiffusion (RID) procedures for IgM are subject to error in the presence of subgroups in IgM.[15]

Good correlation has been shown between kinetic and fixed time nephelometric measurements and, in general, between nephelometric and radial immunodiffusion (RID) procedures.[7] Discrepancies in calibrating standards and batch-to-batch variation have, however, resulted in poor correlation between radial immunodiffusion and nephelometry for assay of specific proteins.[16]

Double diffusion. Double diffusion involves passive diffusion of antigen and antibody toward each other in agar or agarose gel matrix. This procedure is one-dimensional when carried out in a test tube or two-dimensional when performed using a flat gel bed. The essential difference between single and double diffusion is that in the former antigen and antibody are in direct contact, whereas in the latter a spacer gel matrix separates the two (Figure 18–3). The classic Ouchterlony procedure is an example of double diffusion.[17]

The Ouchterlony procedure is usually performed in a Petri dish that has central and peripheral wells separated by the gel matrix. Double diffusion is used primarily to assess the specificity of antigen-antibody systems. Antibody or antigen can be added to the central well, while the peripheral wells are filled with serial dilutions of the corresponding counterpart. As antigen and antibody diffuse toward each other, a precipitin line will be formed provided antigen excess conditions do not prevail. The serial dilutions in peripheral wells are used to overcome the antigen excess

Figure 18–4

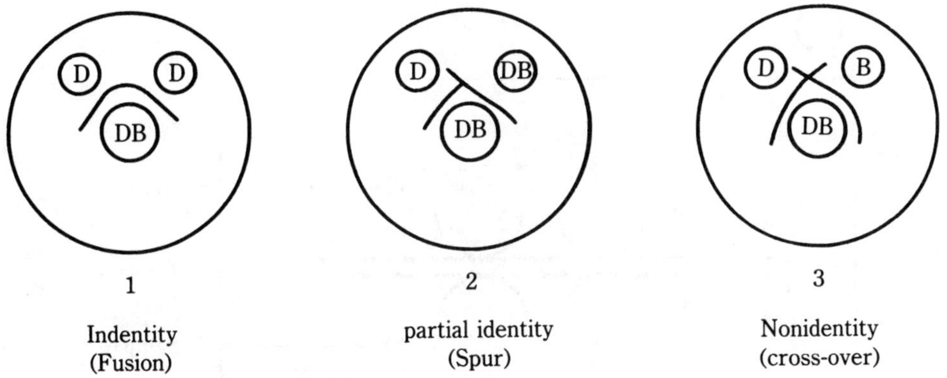

Characterization of the purity of the antigen. The central well of ouchterlony plate contains antibody to digoxin and bovine serum albumin (DB). The peripheral wells contain antigen, either digoxin (D) or bovine serum albumin (B), or a mixture of digoxin-bovine serum albumin. In 1, fusion of immunoprecipitation arc indicates identity since pure D is placed in both peripheral wells. In 2, a spur indicates partial identity since DB is partially identical to D present in the other peripheral well. In 3, the immunoprecipitin arcs cross over, indicating nonidentity since D and B in the respective peripheral wells are not identical.

---

or prozone phenomena. Characteristic immunoprecipitin patterns allow a distinction between antigen-antibody systems of partial, complete and nonidentity. Fusion of immunoprecipitin lines indicates complete identity of antigen and antibody present in the peripheral and central wells respectively. The presence of a spur indicates partial identity. The cross-over of precipitin lines indicates nonidentity (Figure 18–4).

Double diffusion is best suited to measuring molecules < 19 S in size. Although precipitin lines may become visible in 3 to 6 hours, it usually takes 18 to 30 hours for optimal development, and 72 to 96 hours may be required for diffusion of trace amounts of immunoreactants. The double diffusion procedure is capable of detecting 0.6 mg/dL of protein.

Immunoelectrophoresis. Immunoelectrophoresis (IEP) is a qualitative tool for identifying specific proteins in serum. The procedure includes:

1. Electrophoresis of serum in a gel matrix (usually agarose)
2. Addition of antisera to whole human serum or specific antisera to a trough cut out in the gel
3. Allowing separated protein and antisera to diffuse toward each other

After washing off the nonprecipitated protein, precipitin arcs can be visualized by staining or by viewing them in bright light against a dark background (Figure 18–5).

Figure 18–5

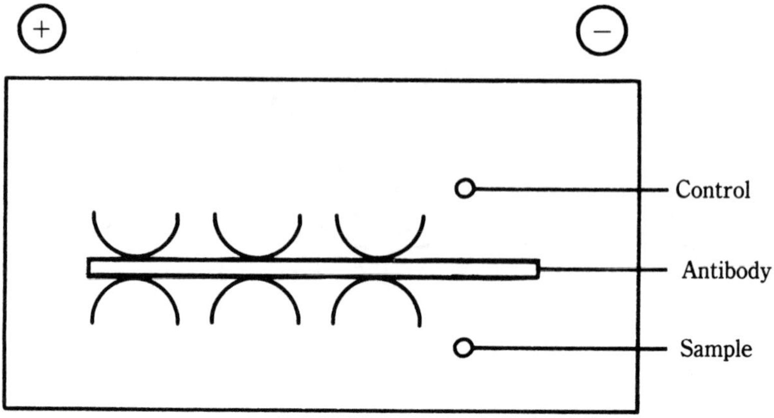

Immunoelectrophoresis. Control containing the same mixture of antigens as the sample is inserted in the control well, while the sample is applied to the sample well of this gel. After electrophoresis, antibody to a mixture of antigens is applied to the antibody trough, and the gel is subsequently incubated in a moist chamber for 24 to 48 hours, or even up to 72 hours. As antigens in the sample and control mixture diffuse toward the antibody, precipitin arcs develop. The control allows indentification of the antigens in the sample by coincidence of precipitin arcs on either side of the antibody trough.

Immunoelectrophoresis (IEP) is essentially electrophoresis followed by immunodiffusion. The procedure is time-consuming; the diffusion step alone takes 24 to 48 hours. Appropriate dilutions of sera are required to ensure that antigen excess does not preclude the formation of certain precipitin arcs. The nature of precipitin arcs by immunoelectrophoresis is used to distinguish monoclonal (spikes) from polyclonal (thick, smooth arcs) gammopathies. Use of agar gel for immunoelectrophoresis may not permit discrimination of small amounts of monoclonal immunoglobulin in the presence of large amounts of polyclonal immunoglobulins of the same class. Use of agarose gel overcomes this limitation.[17]

Counter immunoelectrophoresis. In counter immunoelectrophoresis, the antigen is placed between the antibody and the cathode in an agar gel. During electrophoresis, the antibody, facilitated by the electroendosmotic flow, migrates toward the cathode, while the antigen migrates toward the anode. An immunoprecipitin line forms at the point of union of the antigen and antibody. This technique was originally applied in the detection of hepatitis B surface antigen and antibody.[18]

Crossed immunoelectrophoresis. This is a two-dimensional electrophoretic procedure. The sample is first separated by conventional zone electrophoresis on an agarose gel (first dimension). The gel containing the separated protein fractions

Figure 18–6

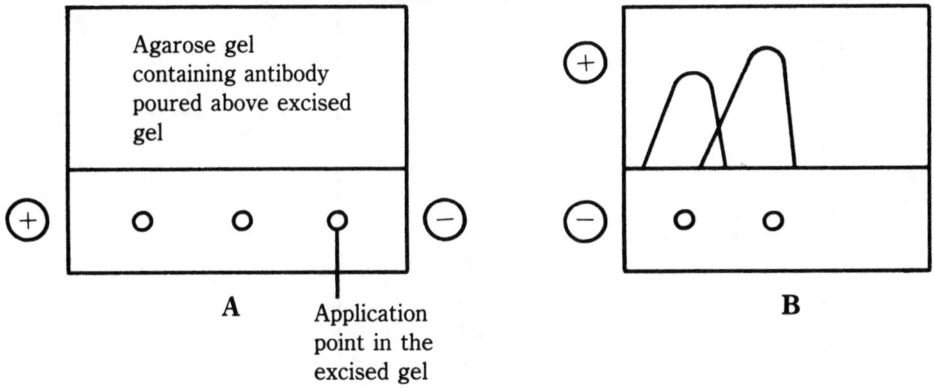

Crossed immunoelectrophoresis. A: Excised gel containing electrophoretically separated fractions is placed on a plate, and agarose-gel–containing polyspecific antiserum is poured above it. B: Electrophoresis carried out 90° to the first run. Separated fractions migrate into antibody-containing gel perpendicular to direction of separation in the first dimension. Area under precipitin arcs allows quantitation of antigen.

is excised and placed on another plate and agarose gel containing polyspecific antiserum is poured above the excised gel. Electrophoresis is now carried out in the second dimension, 90° to the first run. This causes the separated protein fractions to migrate into the polyspecific antiserum containing agarose gel perpendicular to the direction of separation in the first dimension (Figure 18–6).

This is a quantitative technique, since the area of the precipitin peak formed in the second dimension can be measured and related to the calibration curve of area in square millimeters obtained with standards. Because polyspecific antisera is used in the second dimension, a number of proteins can be quantitated in a single run.

Electroimmunodiffusion. Electroimmunodiffusion (EID) employs active diffusion of antigen during electrophoresis on a gel matrix (usually agarose) on which the antibody is immobilized.[19] The precipitin arcs produced in this procedure are shaped like rockets, and the height of the precipitin arc can be related to the concentration of antigen. The procedure is rapid, taking only 2 to 3 hours to complete, as compared with the time consuming radial immunodiffusion procedure (Figure 18–7).

Immunofixation. Immunofixation electrophoresis involves electrophoresis in agarose gel, subsequent layering of antiserum on the agarose gel containing the separated proteins, and incubation in a moist chamber. After washing off the nonprecipitated proteins, the gel is stained to visualize the bands of immune complexes

Figure 18–7

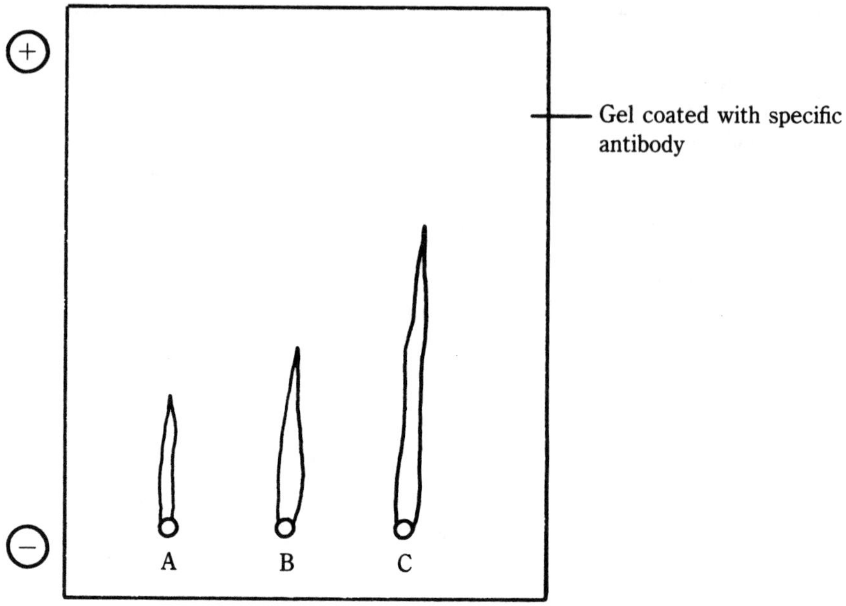

Electroimmunodiffusion. A, B, and C are troughs created in antibody-coated gel for insertion of sample containing specific antigen. Antigen is driven into antibody-coated gel during electrophoresis. The height of the immunoprecipitin arc, which is shaped like a rocket, can be related to the concentration of a specifc antigen in the sample.

trapped or fixed in the gel.[20] The procedure has been useful in the rapid identification of genetic variants.

Standardization of immunoglobulin assays. Since immunoglobulins have been measured by nephelometric, immunodiffusion, and related procedures it is appropriate to address briefly the subject of standardization of immunoglobulin assays. Standardization of immunoglobulin assays has presented problems because of the extreme molecular heterogeneity of the immunoglobulins.[21] The World Health Organization (WHO) has recommended calibration of immunoglobulin standards against an international reference preparation of freeze-dried human serum immunoglobulins (IgG, IgA, and IgM) to which international units have been assigned.[22] Mass units (mg/dL) can then be determined by using appropriate conversion factors.[23] The initial WHO Standard "67/99" was not suitable for use with nephelometry because of its turbidity. To circumvent this limitation, the International Federation for Clinical Chemistry (IFCC) issued a sterile liquid serum preparation "74/1," which is suitable for both nephelometric and immunodiffusion procedures.[24] It should be noted, however, that there are method-specific variations that affect the values established for calibrators.[24] With the many kits available for immunoglobulin measurements, it is imperative that standards or calibrators supplied with such kits are calibrated with reference WHO or IFCC preparations. Even with the use of standards calibrated against an established reference preparation, the presence of monoclonal IgG (M proteins) will result in a high value by radial immunodiffusion. This is caused by differences in the diffusion characteristics of the monoclonal and the heterogeneous immunoglobulin standard. Method-comparison studies on immunoglobulins have been reviewed.[25]

Techniques Utilizing an Indicator Label

Radioassays

Radioassays are essentially competitive binding assays. The two types of competitive binding radioassays that have been utilized are equilibrium and sequential saturation. In the equilibrium assay, both the radioactive labelled antigen and the unlabelled antigen in the serum sample are allowed to compete simultaneously for binding sites on the specific antibody until equilibrium, as defined by the law of mass action, is reached. At this point, the antigen (both labelled and unlabelled) in combination with the antibody is separated from the free labelled and unlabelled antigen. The ratio of bound label to the free labelled antigen is inversely proportional to the concentration of unlabelled antigen in the serum sample.

Several types of equilibrium assays are possible, depending upon the concentrations of antiserum and radioactive label used in the assay. At least three such types have been discussed.[26] In the first type, the dilution of antiserum chosen will bind 33% of the radioactive labelled antigen in the absence of unlabelled antigen. In the second type, conditions are chosen to effect 50% binding of the radioactive label (B/F = 1) to the antibody. In the third case, the antibody is 95% saturated with radiolabelled antigen in the absence of unlabelled antigen. In order to set up these experimental conditions, one must know the affinity constant K of the antibody. The antigen-antibody reaction can be represented by the following equation:

$$\text{Antigen (Ag)} + \text{Antibody (Ab)} \underset{k_2}{\overset{k_1}{\rightleftharpoons}} \text{Antigen-antibody complex (Ag} \bullet \text{Ab)}$$

In this equation, k_1 is the rate constant for the formation of Ag • Ab complex (the forward reaction) and k_2 is the rate constant for the dissociation of the Ag • Ab complex (the reverse reaction). The affinity constant is a ratio of the rates of the reverse and forward reactions (k_2/k_1) at equilibrium. An ideal antiserum should have a very high affinity constant, since the reciprocal of the affinity constant (1/K) of the antiserum is a measure of the sensitivity of the assay. Thus, antisera with an affinity constant of 10^{12} L/mol can measure 10^{-12} mol/L of antigen in sera, thereby achieving a sensitivity of the order of picomoles.

The affinity constant of an antisera can be determined by adding increasing amounts of labelled antigen to a fixed amount of antibody and separating the label bound to the antibody from the free label. Two graphic approaches are available to determine the affinity constant from the data obtained in the above experiment. In one such approach, a Scatchard Plot is used in which the B/F ratio is plotted vs the concentration of the labelled antigen. Such a plot will be linear provided that all binding sites of the antibody have identical affinity for the labelled antigen. Because the slope of this plot will be equal to $-K$, the affinity constant can be readily calculated.[26]

In the second approach, the amount of radiolabelled ligand bound in each experiment is plotted against the concentration of total labelled ligand added to the fixed concentration of antibody. This yields a hyperbolic curve, similar to the one that results when enzyme activity is plotted against increasing substrate concentrations using fixed concentration (activity) of the enzyme. The amount of labelled ligand bound at full saturation of antibody is determined in a separate experiment. The amount of free labelled ligand at 50% saturation of antibody is determined from the graphical plot. The affinity constant is given by the reciprocal of the concentration of free labelled ligand at 50% saturation of antibody. This saturation method of Odell has similarities to the method of determining the Michaelis constant K_m by plotting enzyme activity vs the substrate concentration.[27]

In the three types of equilibrium approaches referred to earlier, the assumption is made that both the labelled and unlabelled antigen have identical affinity for binding sites on the antibody, which permits true competition. The equilibrium approach, where 33% of the radioactive label is bound to the antibody (B/F = 0.5) in the absence of unlabeled antigen, has practical limitations because of its poor precision. The most widely used equilibrium approach is where the B/F = 1 in the absence of unlabelled antigen. This approach is useful for detecting low concentrations of antigen, such as growth hormone and ACTH, provided that the antibody chosen for the assay has a very high affinity constant.

The radioactive label chosen for such assays should have a very high specific activity, such as ^{125}I. In contrast, the third type of equilibrium approach, in which the antibody is 95% saturated with labelled antigen, can be used with antibodies of lower affinity constants, higher amounts of circulating antigen, and a lower specific activity label, such as tritium.[26]

The equilibrium assay cannot be performed if the labelled and unlabelled antigens have different affinities for the antibody. In such cases, the sequential saturation approach might be applicable. In sequential saturation, the unlabelled antigen is allowed to quickly saturate the binding sites on the antibody (within 3 to 4 minutes). Labelled antigen is subsequently added, and the free label is quickly separated from label bound to the antibody. A requirement for this assay is that the primary reaction of the unlabelled antigen with the antibody is essentially irreversible. The amount of radioactive label taken up by the antibody is inversely proportional to the amount of antigen in serum. Because what is measured is the amount of binding sites left over after the initial saturation by the unlabelled antigen, this approach can be used when the radiolabelled and unlabelled antigen have different affinities for the antiserum.[28]

The majority of the competitive binding radioassays are of the equilibrium type in which conditions are chosen to give a B/F ratio of 1 in the absence of unlabelled antigen. In addition to its utility for measuring antigens that have different affinities for the antibody than the radioactive label used for the assay, sequential saturation analysis is suited for measuring analytes, which are found in a reasonably high concentration in serum. In addition, these assays give linearized standard curves without the necessity of resorting to logit-log transformation as is required with the commonly used equilibrium assay.[28]

Separation of bound label from free label. A variety of approaches are available. The addition of Dextran-coated charcoal helps separate the label bound to the antibody that is not retained by the charcoal from the free label captured by the charcoal. Centrifugation leaves the label bound to antibody in the supernatant.

In the double antibody technique, a second antibody raised in a higher animal (assuming the primary was raised in rabbit), such as a goat, is added to the soluble antigen-antibody complex. The secondary antigen-antibody complexes, formed when the second antibody is added, precipitates the label bound to antibody. Centrifugation of the mixture leaves the free label in the supernatant. Either the free or bound label or both may be counted.

Alternatively, if the antibody is coated on a solid phase such as a plastic tube, separation of bound label from free label can be readily accomplished by decanting the incubation mixture and washing the tube. The decanted mixture and the washing will contain the free label, while the label bound to the antibody will stay immobilized on the tube surface.

Labels in radioimmunoassay (RIA). Radioactive decay is a process by which an unstable atomic nucleus attempts to acquire a stable configuration. Radio nuclides are isotopes with the same atomic number (number of protons), but with different atomic or mass number (number of nucleons, which is a sum of both protons and neutrons). Radioactive decay can occur by emission of an α-particle or helium nucleus ($^4_2$He), and such nuclei are called α-emitters.

The decay of radium-226 to yield radon-222 by loss of an α-particle is an example of α-emission. α-emitting radioactive isotopes are nuclides of heavy elements and have limited usefulness in clinical chemistry. Alternatively, an unstable radioactive nucleus can decay without a change in the mass number (atomic weight), but with a change in the atomic number. This occurs when the nucleus emits either a negative or positive beta (β) particle. The emission of a negative β particle ($\beta-$) or negatron confers on the emitting nucleus an additional positive charge that is equivalent to the conversion of a neutron to a proton. Hence, β-emission results in a nucleus with the next higher atomic number. Conversely, emission of a positive β particle ($\beta+$) or positron is equivalent to the conversion of a proton to a neutron and thus results in a nucleus with the next lower atomic number. Examples of β-emitters are Tritium (^{3}H), Carbon-14 (^{14}C), and Phosphorus (^{32}P).

The unstable atomic nucleus may also decay by capture of an electron from orbitals close to the nucleus such as the K and L shells. Decay by electron capture is analogous to $\beta+$ emission, and in both cases proton is converted to a neutron, thus decreasing the atomic number by one. The vacancy created in the inner S orbital as a result of electron capture is quickly filled by an electron from higher orbit. As an electron from a higher orbital falls to the inner orbital vacancy, energy is released in the form of an x-ray. Alternatively, the energy released from the orbital jump may be absorbed with the displacement of a weakly bound orbital electron called the Auger electron.

The nucleus, after either alpha or beta emission or electron capture, may still be in an excited state. Such an unstable nucleus releases its excess energy in the form of electromagnetic radiation of very short wavelength, the gamma (γ) rays. A radionuclide such as $^{125}_{53}I$ first decays by electron capture to metastable $^{125}_{53}Te$ (Tellurium), which further decays by γ emission to stable $^{125}_{53}Te$. γ emission causes no change in either the atomic number or mass number.

Both γ and β emissions can cause ionization and excitation because of their charge and the speed with which they interact with matter. α-particles have a very high ionizing potential when compared to β emitters. The latter, however, have a better penetrating power than α-particles. The uncharged γ-rays, as they are absorbed, can cause the displacement of an electron, provided that the energy of the γ-ray photon is equal to the binding energy of the displaced electron. This mechanism is called the photoelectric effect, and the energy of the photoelectron is equal to the energy of the γ-ray photon.

Alternatively, the γ-ray photon may collide with an electron, thus losing some of its energy to it. This phenomenon is called the Compton Effect, and the resulting photon has less energy than the original γ-ray photon because of the scattering of energy due to collision with electron. γ-rays, although they have low ionizing power, have high penetrating properties. Examples of γ-emitters are Iodine-125, Iodine-131, and Cobalt-58.

The half-life of radionuclides varies from days to several years. Thus, ^{125}I has a half-life of 60 days, ^{131}I a half-life of 8.1 days, ^{58}Co a half-life of 71.3 days, ^{57}Co a half-life of 270 days, ^{3}H a half-life of 12.3 years and ^{14}C a half-life of 5,730 years. ^{125}I, ^{131}I, ^{57}Co and ^{3}H have been used as labels in RIA. The choice of label is dictated by both its half-life and also its specific activity. Specific activity is the activity of the radionuclide per weight of the sample and is expressed as μci (microcuries) per μg of radionuclide.

The higher the specific activity, the greater the sensitivity of the label used in RIA. Thus, ^{3}H with a specific activity of 9.7 μci/μg has a much lower specific activity than ^{125}I with a specific activity of 17.3 μci/μg and is, therefore, a less sensitive label than ^{125}I. ^{131}I has a very high specific activity of 123 μci/μg. However, its half-life is very short (8.1 days) compared to ^{125}I with a half-life of 60 days.

Scintillation counters. Solid scintillation counters are used to count γ radiation. A sodium iodide crystal activated with 1% thallium is used as the solid scintillator. The crystal is in the form of a well to allow the introduction of sample to be counted. The crystal is sealed onto a photomultiplier tube to detect light pulses emitted by the crystal at a wavelength of 410 nm in less than a microsecond interval. Because the sodium iodide crystal is highly hygroscopic, it has to be encased in an aluminum foil to prevent it from absorbing moisture except in the area where the photomultiplier tube is facing the crystal, which is covered by a transparent material.

The highly penetrating γ-rays from the radioactive label in the sample strike the sodium iodide crystal, and photons are released as the radiation energy is absorbed by the crystal. The photons are detected by the photomultiplier tube, which amplifies the signal to produce a cascade of electrons. The output from the photomultiplier tube can be fed to a pre-amplifier, which is further amplified by another amplifier prior to being fed to a pulse height analyzer and scaler or recording

Figure 18–8

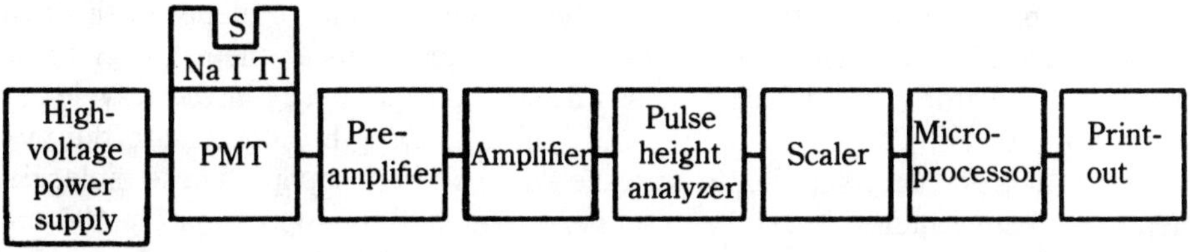

Design of a solid scintillation counter. S = the well where sample tube is inserted. Na I T1 = the sodium iodide crystal activated with 1% thallium, which is the solid scintillator. PMT = the photomultiplier tube.

device. Figure 18–8 is a schematic representation of a solid crystal scintillation counter.

The pulse height analyzer sets the window width for counting. Because each radionuclide has a characteristic energy spectrum for pulse height, the pulse height analyzer can define the window width by setting lower and upper discriminator levels. Pulses below the lower discriminator levels are not counted. Only those pulses defined by the window width and constituting the photo peak are thus counted. These pulses are processed through an anticoincident circuit, which corrects for the probability of two oscillations from both the upper and lower discriminators arriving at the same time, prior to their being fed to the scaler. The scaler counts the number of pulses arrived in a defined time interval. The output from the scaler can be processed by a microprocessor to display data in chart form and perform calculations to yield quantitative results on analytes in the sample.

Multichannel analyzers capable of counting two or more radionuclides simultaneously utilize separate pulse height analyzers for each radionuclide. As long as the photopeaks of two or more radionuclides do not sufficiently overlap, it is possible to detect a mixture of radiolabels simultaneously in the same sample. Thus, ^{131}I, which has a major photopeak of 364 Kev, can be counted without interference from ^{125}I major photopeak at 27 Kev. However, since ^{131}I has a minor photopeak at 32 Kev, it would interfere in the window width of 20 to 40 Kev chosen in the ^{125}I channel. This can be overcome by estimating the amount of ^{131}I contribution to the ^{125}I channel by counting an ^{131}I standard is both the ^{131}I and ^{125}I channels.

Solid scintillation counters can be used to count radionuclide labels, such as ^{125}I, ^{131}I, and ^{57}Co.

The solid scintillation counters are limited in their ability to detect β-emission, because they cannot penetrate the sample tube wall and the aluminum foil encasing the solid sodium iodide crystal.

In a liquid scintillation counter, the sample containing the β-emitting radionuclide such as ^{3}H, ^{14}C, or ^{32}P is dissolved in a scintillation cocktail. A typical cocktail consists of a mixture of an organic solvent such as toluene, a primary scintillator such as 2,5-diphenyloxazole (ppo), and a secondary scintillator or wavelength shifter, such as 1,4-Di-2-(5phenyloxazolyl)-benzene (popop), and solubilizers and surfactants to maintain homogeneity of sample and additives to improve counting efficiency.

The β-particles in the sample ionize the organic solvent (toluene) in the scintillation cocktail. The excited molecules of the solvent transfer their energy to the primary scintillator (ppo), which releases their acquired energy in the wavelength of 350 to 400 nm. The photons emitted in this wavelength range excite the molecules of the secondary scintillator or wavelength shifter (popop). These molecules then emit their acquired energy at a longer wavelength of approximately 430 nm, which can be readily detected by the photomultiplier tubes. The design features of the liquid scintillation counter are essentially similar to those of the solid scintillation counter, except that the scintillation cocktail is placed between two photomultiplier tubes. By comparing pulses from the two photomultiplier tubes by means of a coincidence circuit, only pulses that correspond to scintillations produced by the radiolabel are processed. Noise pulses, whose probabilities of being produced simultaneously by both photomultiplier tubes are relatively small, are excluded.

Problems in liquid scintillation counting include quenching by impurities which absorbs the emitting photons, and thus reduces the intensity of emission and side reaction such as chemiluminescence. The latter results from interaction of sample with one of the additives in the scintillation cocktail. The quenching phenomena can be corrected by the determination of efficiency, which is the ratio of observed counts per minute to the expected disintegrations per minute (dpm) for the radionuclide. Correction for quenching can be handled by a variety of approaches with liquid scintillation counters, which includes monitoring of counts at two different window settings, and the inclusion of both internal and external standards.

Nonisotopic Immunoassays

The use of a nonisotopic label is appealing in that it dispenses with the concerns of handling radioactive materials, even though the amount of radioactivity of the order of 0.1 μci encountered in RIA procedures is too small to pose a serious health hazard.

Both enzyme and fluorescence labels have been successfully used in such assays, either with or without separation steps. Where no separation step is involved, the nonisotopic immunoassay is considered to be a homogeneous assay. When separation or washing steps are required, as in RIA procedures, the assay is considered to be an heterogeneous assay.

Homogeneous enzyme immunoassay. In a widely used homogeneous enzyme immunoassay, the enzyme-linked antigen is allowed to compete with the antigen in the sample for binding sites on the specific antibody. In the absence of antigen in the sample, the substrate for the enzyme is sterically excluded from the active site of the enzyme, thus preventing formation of the product. In the presence of antigen in the sample, however, competition develops between it and the enzyme-linked antigen for the binding sites on the antibody, thus altering the conformation of the enzyme to permit substrate attachment to the active site. The amount of substrate admitted to the active site of the enzyme, and consequently the amount of product formed, is directly proportional to the concentration of antigen in the sample. Thus, by measuring a property of the product formed, such as absorbance in the ultraviolet region or disappearance of turbidity in a spectrophotometer, one can assess the concentration of antigen in the sample. Homogeneous enzyme immunoassays can also be set up to utilize inhibition of product formation to assess the concentration of antigen in the sample. Lysozyme and glucose-6-phosphate dehydrogenase are examples of some of the enzymes that have been used as labels in homogeneous enzyme immunoassays.[29,30]

Heterogeneous enzyme immunoassays. Enzyme-linked immunosorbent assay (ELISA) typifies heterogeneous enzyme-immunoassays. Essentially, these are solid phase assays with the antigen or the antibody bound to a solid phase. The three types of such assays that are commonly in use are: the double antibody sandwich technique for measurement of antigen; the competitive binding assay for detection of antigen; and, the indirect assay for the determination of levels of antibody. In the double antibody sandwich assay, antigen present in the sample is allowed to react with antibody adsorbed to a solid phase. After a washing step to remove the unadsorbed species, an enzyme-labelled specific antibody is added. Excess of enzyme-labelled antibody remaining after the formation of a sandwich is washed away. Substrate for the enzyme is then added, and the amount of product formed is measured in a spectrophotometer and is directly proportional to the concentration of antigen in the sample.

The competitive binding assay is performed both in the presence and absence of enzyme-labelled antigen added to antibody adsorbed to a solid phase. After a washing step to remove antigen, substrate is added. The absorbance of the product formed in the presence of enzyme-linked antigen and antigen in the sample is subtracted from the absorbance of the product formed in the presence of enzyme-linked antigen alone. The difference in absorbance can be related to the concentration of antigen in the sample. In the indirect assay, the antigen is adsorbed to the solid phase and the sample containing specific antibody is added. After washing to remove unreacted species, enzyme-labelled antihuman globulin is added. Any remaining free enzyme label is subsequently washed away before addition of substrate. The absorbance of the product formed is directly related to the concentration of antibody in the sample. Alkaline phosphatase and horseradish peroxidase are examples of some of the many enzymes that have been used as labels in enzyme-linked immunosorbent assays.[31] Competitive enzyme-linked immunoassays have been used for the measurement of IgG.

Nonhomogeneous (heterogeneous) fluoroimmunoassays. A few representative types of the various enzyme immunoassays that use fluorescence as a label will be considered. A nonhomogeneous fluoroimmunoassay for IgG utilizes the initial reaction of serum with fluorescein-isothiocyanate-conjugated antibody to IgG. The unreacted fluorescent antibody is taken up on a polymeric surface coated with IgG. The fluorescence adsorbed onto this solid phase is quantitated in a fluorometer. Since the amount of unreacted fluorescent antibody is inversely proportional to the concentration of antigen in the sample, the amount of fluorescence bound to the solid phase will also be inversely proportional to the concentration of the antigen in the sample.

Another nonhomogeneous approach to the measurement of immunoglobulins utilizes specific antibodies bound to derivatized polyacrylamide beads. After appropriate incubation of sample containing a specific immunoglobulin with the specific antibody-bound solid phase and after subsequent washing, fluorescent-labelled specific antibody is added. After another washing step the amount of fluorescence bound to the antigen is directly related to its concentration. These two types of fluoroimmunoassays for the measurement of immunoglobulins in serum have been compared with a radial immunodiffusion procedure.[32] IgG values obtained by these fluoroimmunoassays are comparable to those obtained with radial immunodiffusion. IgM values obtained with these fluoroimmunoassays are significantly lower, however, than the corresponding values obtained with radial immunodiffusion. IgA values obtained with the first of the fluoroimmunoassays discussed above are in close agreement with the values obtained by radial immunodiffusion. Fluoroimmunoassays using derivatized polyacrylamide beads, however, give significantly higher IgA values compared to those obtained by radial immunodiffusion. The diffusion characteristics of monomeric IgM and polymeric IgA might help explain the discrepancy between radial immunodiffusion and other assays where diffusion is not a limiting factor.

Homogeneous fluoroimmunoassays. An example of a homogeneous fluoroimmunoassay is the fluorescence protection immunoassay. In a typical procedure designed for measurement of IgG, serum anti-human IgG (antibody) and the fluorescein-labelled human IgG are incubated at room temperature in a test tube for 30 minutes. Anti-fluorescein is subsequently added and the reaction mixture is incubated for 30 minutes. Anti-fluorescein is unable to quench the fluorescence of fluorescent-labelled IgG bound to anti-human IgG (antibody) because it is sterically excluded from the bound fluorescein label. The free fluorescent-labelled IgG, not in combination with antibody, is quenched by the anti-fluorescein, and this decrease in fluorescence is directly related to the concentration of IgG in the sample. Rate of fluorescence quenching can be monitored during a 30 to 60 second interval, thereby obviating the use of serum blanks, since serum background fluorescence is automatically subtracted. The homogeneous fluorescence protection assay for IgG has shown good correlation when compared with radial immunodiffusion.[33]

Another example of a homogeneous fluoroimmunoassay is the fluorescent excitation transfer immunoassay.[34] In such an assay, the antibody is labelled with an acceptor molecule such as tetramethyl rhodamine, while the antigen is labelled with a fluorescer such as fluorescein. The labelled antibody, the labelled antigen, and the antigen in the serum sample are incubated together. If there is less antigen in the serum sample, more of the labelled antigen will bind to the labelled antibody. This results in the transfer of fluorescence energy from the labelled fluorescent antigen to the acceptor molecule on the antibody, leading to fluorescence quenching. Thus, the measurement of fluorescence caused by the free fraction of the labelled antigen, (whose fluorescence is not quenched, since it does not bind to the antibody), can be inversely related to the concentration of antigen in the serum sample.

The substrate-labelled fluorescence immunoassay is another example of nonseparation or homogeneous fluoroimmunoassay.[35] This assay has found wide application in the measurement of drug levels in serum. In this assay, the antigen bound to a nonfluorescent molecule, such as umbelliferyl-β-galactoside serves as the label. The assay is based on the competition of antigen in the serum sample and the labelled antigen for binding sites on the antibody.

In this assay, an enzyme such as β-galactosidase that hydrolyses the antigen-labelled umbelliferyl-β-galactoside molecule to yield the fluorescent product umbelliferone is added. The labelled antigen bound to the antibody is sterically excluded from hydrolysis by the enzyme. The free-labelled antigen, which is not bound to the antibody, is hydrolyzed, resulting in the release of the fluorophore (umbelliferone). The increase in the measured fluorescence is thus directly related to the concentration of antigen in the serum sample.

In addition to the fluoroimmunoassays discussed above, several different approaches to fluoroimmunoassays have been described. One novel approach utilizes the measurement of fluorescence polarization with an instrument that has found wide application.[36] This approach is based on the principle that a fluorescent molecule excited by polarized light will emit light in the plane of polarization. The polarization of the emitted light is related to the Brownian movement of the molecules during the short lifetime of the excited state, which is of the order of 4 to 5 nanoseconds for fluorescein-labelled molecules. The smaller the molecule, the faster it will tumble, and the less will be the amount of emitted light in the plane of polarization. In other words, fluorescence will be depolarized.

Fluorescent molecules can absorb polarized light only if they are oriented in such a manner that their dipoles or absorption oscillators are parallel to the plane of vibration of the exciting light. Such molecules also emit polarized light in the same plane.

Fluorescence polarization is related to the rotational relaxation time of the molecule and the fluorescence lifetime of the molecule. Rotational relaxation time of the molecule is the time required for a properly oriented molecule to assume a random orientation when it is excited with polarized light. The fluorescence lifetime of the molecule is the time interval between excitation of molecule and the time the molecule emits the absorbed light at a longer wavelength.

Fluorescence polarization finds application in immunoassay because small molecules have short rotational relaxation times of the order of 1 nanosecond, reflecting the rapidity with which they rotate or tumble in solution. Therefore, they will be randomly oriented within the relatively long average fluorescence lifetime of a molecule, which ranges from 10^{-8} to 10^{-9} seconds. For this reason, small molecules, by the time they emit the absorbed polarized light, will be in a random orientation and will be unable to emit polarized light. In contrast, large molecules, such as labelled drug or hapten bound to antibody, because of their long rotational relaxation times of approximately 100 nanoseconds, will rotate or tumble slowly throughout the fluorescence lifetime of the molecule. As such, the large molecule will have its dipoles properly oriented to emit polarized light.

The instrument used to measure fluorescence polarization is essentially a fluorometer with a tungsten-halogen light source. The wavelength of the primary excitation filter is 485 nm since fluorescein is used as a label. This excitation wavelength is passed through a polarizer, which transmits vertically polarized light that is then passed through a liquid crystal. Under the influence of an electric current, the liquid crystal is able to rotate the vertically polarized light 90° to alternate the plane of polarization from vertical to horizontal.

The radiation from the liquid crystal passes through the sample cuvette where the fluorescein-labelled molecules are excited alternately by biplanar (vertically and horizontally polarized) radiation. The resulting fluorescence emission in the range of 525 to 550 nm is isolated with an interference filter, the secondary emission filter, which is positioned at 75° angle from the cuvette. The light emerging from the emission filter passes through a polarizer that transmits fluorescence emission vertically polarized onto the detector, which is a photomultiplier tube.

Fluorescence polarization or the degree of polarization (P) can be calculated by knowing the detector response when the sample is excited with vertically polarized light (l_{vv}), and the detector response to vertically polarized emission resulting from excitation of sample with horizontally polarized light (I_{hv}). The degree of polarization is calculated by the instrument's microprocessor with the following formula:

$$P = \frac{I_{vv} - I_{hv}}{I_{vv} + I_{hv}}$$

I_{hv} is essentially a measure of free labelled drug, whereas I_{vv} approximates the amount of label bound to antibody. In fluorescence polarization immunoassay, which has been widely used for the assay of drugs, a fluorescein-labelled drug (small molecule), and the drug in serum compete for binding sites on the antibody. Because the labelled drug tumbles rapidly, its fluorescence will be depolarized. The degree of polarization is thus directly related to the amount of labelled drug bound to antibody (large molecule).

Because the concentration of drug in serum will dictate how much labelled drug will bind to antibody, the increase in the degree of fluorescence polarization is inversely related to the concentration of the drug in serum. By knowing the degree of fluorescence polarization of calibrating standards, the degree of fluorescence polarization of an unknown sample can be automatically related to the drug concentration.

Fluorescence polarization immunoassay has become an established technique in the therapeutic drug monitoring and toxicology laboratory.

Fluoroimmunoassays offer a choice of labels that include fluorescein, tetramethylrhodamine, umbelliferone, phycobiliproteins, and rare earth chelates such as Europium chelates.

Although fluorescein, with its high quantum yield, ease of labelling, and stability has been widely used in both fluoroimmunoassays and immunofluorometric assays (antibody is labelled), the difference in the excitation (490 nm) and fluorescence emission wavelength (520 nm) is narrow, thus dictating the use of filters with a narrow band pass. In contrast, the rare earth chelates have a 200 nm difference between the excitation and emission wavelength.

Phycobiliproteins isolated from algae are stable and possess high quantum efficiency. Their excitation and emission wavelengths are in the red end of the spectrum, permitting use of low-energy light sources such as tungsten lamps. These compounds are hydrophilic and can be readily conjugated to proteins such as IgG, thus allowing the design of sensitive fluoroimmunoassays.

Applications

Such rapid advances are made in the field of immunoassays that it would be impossible to discuss the many variations. However, a few of the recent applications deserve mention.

The availability of monoclonal antibodies directed to separate sites on the antigen molecule has permitted the development of specific two-site immunometric procedures. Thus the measurement of human chorionic gonadotropin (HCG) in serum or urine can be performed in disposable membrane devices in which antibody to the alpha subunit of HCG is localized in the center of the membrane.[37] It is necessary only to add serum or urine (500 μl to 250 μl respectively) and 3 drops of antibody directed to the beta subunit of HCG and conjugated to an enzyme such as alkaline phosphatase. The antibody enzyme conjugate will bind only to the HCG immobilized by the antibody on the membranes. After a minute's incubation, the membrane is drained with buffer, 3 drops of substrate such as indoxyl phosphate are added, and product formation is allowed to continue for 2 minutes for urine samples and 3 minutes for serum samples. The reaction is then stopped with wash solution. The appearance of indigo blue in the center of the membrane indicates that HCG was present in the sample. The device can detect as little as 25 IU/L of HCG in serum and 50 IU/L of HCG in urine. The above membrane format can also incorporate a reference zone to which antibody to alkaline phosphatase is immobilized. The reference zone can also serve as an internal reference beside the zone where antibody directed to the alpha subunit of HCG is incorporated.[38] The incorporation of an internal reference zone compensates for variations in assay conditions and for the presence, if any, of interfering substances that will affect both the reference and the test zones to the same extent. A variation of two-site immunoassay is the so-called particle concentration fluorescence immunoassay (PCFIA).[39] In this assay, polystyrene latex particles coated with specific antibody and fluorescein-labelled specific antibody are added to an assay plate or well to which serum has been added. If IgG is to be measured by this procedure, the reagents that would be added to the serum sample would be antibody to IgG and fluorescein-labelled goat anti-human IgG. IgG in the sample would combine with the latex particle coated with antibody to IgG, which would bind with fluorescein-labelled goat antihuman IgG to form a sandwich. In an instrument that utilizes this approach and has 96-well assay plates, each plate with serum and the respective reagents are incubated for 17 minutes at room temperature. Subsequently, the wells are washed through a filter at the base of each plate or well by applying vacuum. The amount of fluorescence label bound to the particle on the surface of the plate is measured by front surface fluorescence and is proportional to the specific antigen concentration in the sample. Inclusion of internal fluorescent reference particles with the antibody-coated particles allows correction for inhomogeneity from well to well and pipetting errors, because a ratio of sample fluorescence to internal reference standard fluorescence compensates for such errors.

A novel immunoassay approach is the utilization of a cofactor for enzyme as analyte-conjugate to activate an inactive enzyme (apoenzyme) and relate the product formed by activated enzyme (holoenzyme) to the concentration of analyte in the sample.[40] Thus inactive glucose oxidase (apoglucose oxidase) can be activated by cofactor flavin adenine dinucleotide (FAD) to produce active glucose oxidase (holoenzyme) that can then act on its substrate glucose and oxygen to produce gluconic acid and hydrogen peroxide. One can measure the amount of oxygen consumed by either using an oxygen electrode (polarography) or coupling it with peroxidase and a substrate in reduced form such as 4-aminoantipyrine to yield an oxidized chromogen that is measured by spectrophotometry. To measure an analyte such as thyroxine-binding globulin (TBG) by this apoenzyme reactivation immunoassay system (ARIS), the reagents used would be antibody to TBG and FAD-labelled TBG. TBG in serum competes with FAD-labelled TBG for binding sites on the antibody to TBG. Apoglucose oxidase is subsequently added to combine with free FAD-labelled TBG (the fraction not bound to the antibody) to produce active glucose oxidase that in turn is measured by the peroxidase-coupled reaction. Because FAD-labelled TBG that is in combination with antibody cannot activate apoglucose oxidase, the amount of product formed by activated glucose oxidase is directly proportional to the amount of TBG in serum.

A simple quantitative immunoassay using whole blood relies upon chromatographic separation. In this elegant assay, monoclonal antibody to a drug such as theophylline and glucose oxidase are immobilized on a paper strip.[41] A few microliters of whole blood are first added to a tube containing a milliliter of reagent mixture consisting of theophylline-labelled horseradish peroxidase, glucose, ascorbate, and dicarboxidine, and mixed. The paper strip containing immobilized theophylline antibody and glucose oxidase is then inserted into the tube, and the reagent mixture rises by capillary action up the paper strip during a 20 minute incubation step. During this period, the theophylline in whole blood competes with horseradish peroxidase-labelled theophylline (HRP-theophylline) for binding sites on the immobilized antibody on the paper strip. HRP-theophylline is displaced upward on the strip by theophylline in whole blood. Thus the height to which HRP-theophylline is bound can be related to the theophylline concentration in whole blood. HRP-theophylline can be visualized as it converts hydrogen peroxide (produced by glucose and glucose oxidase) and dicarboxidine to a colored product. At the completion of the incubation step, the height of the colored band is measured, and the theophylline concentration in the sample is computed from a standard curve relating migration band height to theophylline concentration. Thus the height of the color band is directly proportional to the concentration of the desired analyte in the sample. The presence of ascorbate in the reagent mixture prevents color formation until after the incubation step is complete, during which all drug in the blood would have been bound by the antibody on the strip. Ascorbate prevents premature color formation by reconverting oxidized colorless intermediates of dicarboxidine produced by reaction of HRP-theophylline and hydrogen peroxide back to dicarboxidine, and in the process ascorbate is converted to dehydroascorbate. It is only after all the ascorbate is consumed, which corresponds to the incubation time, that oxidized colorless intermediates (produced by action of HRP-theophylline and hydrogen peroxide or dicarboxidine) are converted to colored product, and the colored band on the paper strip becomes visible. This enzyme immunochromatography format is ideally suited to performing therapeutic drug monitoring at the bedside or in the doctor's office without the aide of any instrumentation.

The ability to enclose reagents in phospholipid vesicles or liposomes has been skillfully exploited in the design of several immunoassays.[42–44] In these assays the entrapped agent released from the liposomes during assay can be related to the analyte concentration in the sample. The liposome can be ruptured in a variety of ways, one of which is lysis by using complement. A homogenous liposome immunoassay that uses complement-mediated lysis has on the exterior of the liposome lipid conjugated drug, and encapsulated within the liposome is the enzyme glucose-6-phosphate dehydrogenase and its substrate glucose-6-phosphate.[43] These liposomes are incubated with serum, antibody to specific drug, complement, and buffer containing cofactors such as NAD+, an indicator dye and an electron acceptor. In the presence of specific drug in serum, the lipid-conjugated drug on the surface of the liposome binds the antibody to the drug. The antibody-lipid conjugated drug complex activates the complement and lyses the liposome to release enzyme and substrate that react with the assay system to generate NADH, which in turn reduces the blue dye (2,6-dichlorophenolindophenol) in presence of electron acceptor (phenazine methosulfate) to a colorless compound. If the specific drug is present in serum, it will compete with lipid-conjugated drug on the surface of the liposome for binding sites on the antibody. This will slow down or inhibit the complement-mediated lyses and in turn decrease the amount of blue dye reduced. Thus the extent of disappearance of blue color at 610 nm is inversely proportional to the concentration of specific drug in serum. Alternatively, one could measure the absorbance of NADH liberated in the assay at 340 nm without resorting to coupling with electron acceptor. However, the fact that both human serum and guinea pig complement absorb at 340 nm contributes to background absorption and thus limits the utility of this approach.

Some of the drawbacks associated with solid-phase–based immunoassays include nonspecific binding to the solid phase, and the limiting rate of diffusion of reactants by the presence of an undisturbed layer of solution at the surface of the solid-phase, the so-called nernst layer. These drawbacks can be overcome if the solid-phase is generated after the specific binding reaction has occurred. In such phase-separation immunoassays, a monoclonal antibody to an analyte, such as IgG conjugated to a monomer (the capture antibody), is incubated with serum, a fluorescent-labelled monoclonal antibody to the gamma-heavy chain of IgG, and a co-monomer compound such as hydroxyethyl methacrylate. A sandwich of capture antibody-IgG in serum-fluorescent–labelled antibody is formed during the incubation step. After the sandwich is formed, the monomer bound capture antibody that is now part of the sandwich is polymerized with co-monomer by addition of free radical polymerization reagents. The fluorophore bound to the polymerized solid-phase can be measured in a microfluorimeter.[45] Alternatively solid-phase can be generated by the use of a capture antibody conjugated to a polymer that is soluble while the sandwich is formed with analyte in serum and fluorescent-labelled antibody. After the sandwich is formed, temperature is raised, rendering the polymer that is now part of the sandwich insoluble. Fluorescence of the generated solid phase is directly proportional to the concentration of specific analyte in serum.

In conclusion, it should be noted that circulating IgG antibodies to nonhuman immunoglobulins (the hetrophile antibodies) can be a source of interference in all immunoassays.[46] Thus in two-site immunometric assays, the binding of heterophile antibodies in human serum to mouse antibodies used as reagents for capture and indicator antibody can mimic the presence of the desired analyte and cause its overestimation and a false-positive result. The effect of heterophile antibody will depend on the size of immune complex that is formed when the reagent antibody combines with heterophile antibody. Thus when heterophile antibodies in serum are in excess, there may even be an underestimation of the desired analyte. Analyte binding to the indicator antibody can also be reduced by heterophile antibodies, in which case the level of the measured analyte is underestimated. The laboratorian should thus be alert to such practical problems in immunoassays.

References

1. Killingsworth LM, Savory J: Automated immunochemical procedures for measurement of immunoglobulins IgG, IgA and IgM in human serum. Clin Chem 17:926, 1971.
2. Ritchie RF, Alper CA, Graves J, et al: Automated quantitation of proteins in serum and other biological fluids. Am J Clin Pathol 59:151, 1973.
3. Tiffany TO, Parella JM, Johnson WF, Burtis CA: Analysis by light scattering measurement with a miniature centrifugal fast analyzer. Clin Chem 20:1055, 1974.
4. Buffone GJ, Savory J, Cross RE, Hammond JE: Evaluation of kinetic light scattering as an approach to the measurement of specific proteins with the centrifugal analyzer. I: Methodology. Clin Chem 21:1731, 1975.
5. Deaton CD, Maxwell KW, Smith RS, Creveling RL: Use of laser nephelometry in the measurement of serum proteins. Clin Chem 22:1465, 1976.
6. Sternberg JC: A rate nephelometer for measuring specific proteins by immunoprecipitin reactions. Clin Chem 23:1456, 1977.
7. Blanchard GC, Gardner R: Two nephelometric methods compared with a radial immunodiffusion method for the measurement of IgG, IgA and IgM. Clin Biochem 13:84, 1980.
8. Walsh RL, Coles ME, Geary TD: Evaluation of two commercially available laser nephelometers and their recommended methods for measurement of serum immunoglobulin G. Clin Chem 25:151, 1979.
9. Oudin J: Specific precipitation in gels and its application to immunochemical analysis. In AC Corcoran (ed): Methods in Medical Research, vol 5. Year Book, Chicago, 1952, p 335.
10. Fahey JL, McKelvey E: Quantitative determination of serum immunoglobulins in antibody-agar plates. J Immunol 94:84, 1965.
11. Mancini G, Carbonara AO, Hermans JF: Immunochemical quantitation of antigens by single radial immunodiffusion. Immunochem 2:235, 1965.
12. Ritzmann SE, Fischer GL, Nakamura RM: Quantitative Immunochemical Procedures. In SE Ritzmann, JC Daniels: Serum Protein Abnormalities, Diagnostic and Clinical Aspects. Little Brown, Boston, 1975, p 61.
13. Virella G, Fudenberg HH: Comparison of immunoglobulin determinations in pathological sera by radial immunodiffusion and laser nephelometry. Clin Chem 23:1925, 1977.
14. Alexander RL, Jr: Comparison of radial immunodiffusion and laser nephelometry for quantitating some serum proteins. Clin Chem 26:314, 1980.

15. de Bruyn AM, Klein F: The validity of the radial immunodiffusion method for the quantitative determination of IgM: Evaluation of a modified method. J Immunol Methods 11:311, 1976.

16. Shulman G: Comparison of specific protein assays in biological fluids by radial immunodiffusion and laser nephelometry. Clin Biochem 12:123, 1979.

17. Smith AM, Thompson RA, Haeney MR: Detection of monoclonal immunoglobulins by immunoelectrophoresis: A possible source of error. J Clin Pathol 33:500, 1980.

18. Hicks EJ, Hughes BJ: Comparative sensitivities of radioimmunoassay, crossover electrophoresis and agar gel diffusion for HB Ag detection. Am J Clin Pathol 65:540, 1976.

19. Laurell CB: Quantitative estimation of proteins by electrophoresis in agarose gel containing antibodies. Anal Biochem 15:45, 1966.

20. Alper CA, Johnson AM: Immunofixation electrophoresis: A technique for the study of protein polymorphism. Vox Sang 17:445, 1969.

21. Reimer CB, Maddison SE: Standardization of human immunoglobulin quantitation: A review of current status and problems. Clin Chem 22:577, 1976.

22. Anderson SG: Measurement of concentrations of human serum immunoglobulins. J Clin Pathol 25:133, 1972.

23. Humphrey JH, Batty IJ: International reference preparation for human serum IgG, IgA and IgM. Clin Exp Immunol 17:714, 1974.

24. Whicher JT, et al: Method-specific variations in the calibration of a new immunoglobulin standard suitable for use in nephelometric techniques. Clin Chem 24:531, 1978.

25. Narayanan S: Method-comparison studies on immunoglobulins. Clin Chem 28:1528, 1982.

26. Zettner A: Principles of competitive binding assays (saturation analysis). I: Equilibrium techniques. Clin Chem 19:699, 1973.

27. Odell WD, Abraham G, Raud HR, et al: Influence of immunization procedures in the titer, affinity, and specificity of antisera to glycopeptides. Acta Endocrinol Suppl. 140:54, 1969.

28. Zettner A, Duly PE: Principles of competitive binding assays (saturation analysis). II: Sequential saturation. Clin Chem 20:5, 1974.

29. Rubenstein K, Schneider R, Ullman E: "Homogeneous" enzyme immunoassay: A new immunochemical technique. Biochem Biophys Res Commun 47:846, 1972.

30. Ullman EF, Blakemore J, Lente RK, Eimstad W, Jaklitsch A: Homogeneous enzyme immunoassay for thyroxine (Abstract). Clin Chem 21:1011, 1975.

31. Schuurs AHWM, Van Weemen BK: Enzyme-Immunoassay. Clin Chem Acta 81:1, 197.

32. Blanchard GC, Gardner R: Two immunofluorescent methods compared with a radial immunodiffusion method for measurement of serum immunoglobulins. Clin Chem 24:808, 1978.

33. Zuk RF, Rowley GL, Ullman EF: Fluorescence protection immunoassay: A new homogeneous assay technique. Clin Chem 25:1554, 1979.

34. Ullman EF, Schwarzberg M, Rubenstein KE: Fluorescent excitation transfer immunoassay: A general method for determination of antigens. J Biol Chem 25:4172, 1976.

35. Burd JF, Carrico RJ, Fetter MC, et al: Specific protein binding reactions monitored by enzymatic hydrolysis of ligand-fluorescent dye conjugates. Anal Biochem 77:56, 1977.

36. Dandliker WB, Kelly RT, Dandliker J, et al: Fluorescence polarization immunoassay: Theory and experimental method. Immunochem 10:219, 1973.

37. Valkirs GE, Barton R: Immunoconcentration—a new format for solid-phase immunoassays. Clin Chem 31:1427, 1985.
38. Anderson RR, Lee TT, Saewert DC, Sowden KM, Valkirs GE: Internally referenced immunoconcentration assays. Clin Chem 32:1692, 1986.
39. MacCrindle C, Schwenzer K, Jolley ME: Particle concentration fluorescence immunoassay: A new immunoassay technique for quantitation of human immunoglobulins in serum. Clin Chem 31:1487, 1985.
40. Schroeder HR, Dean CL, Johnson PK, Morris DL, Hurtle RL: Coupling aminohexyl-FAD to proteins with dimethyladipimidate. Clin Chem 31:1432, 1985.
41. Chen R, Li TM, Merrick H, et al: An internal clock reaction used in a one-step enzyme immunochromatographic assay of theophylline in whole blood. Clin Chem 33:1521, 1987.
42. O'Connell JP, Campbell RL, Fleming BM, et al: A highly sensitive immunoassay system involving antibody-coated tubes and liposome-entrapped dye. Clin Chem 31:1424, 1985.
43. Canova-Davis E, Redemann CT, Vollmer YP, Kung VT: Use of a reversed-phase evaporation vesicle formulation for a homogenous liposome immunoassay. Clin Chem 32:1687, 1986.
44. Ullman EF, Tarnowski T, Felgner P, Gibbons I: Use of liposome encapsulation in a combined single-liquid reagent for homogenous enzyme immunoassay. Clin Chem 33:1579, 1987.
45. Auditore-Hargreaves K, Houghton RL, Monji N, et al: Phase separation immunoassays. Clin Chem 33:1509, 1987.
46. Boscato LM, Stuart MC: Hetrophilic antibodies: A problem for all immunoassays. Clin Chem 34:27, 1988.

19
DNA Probes

The fact that instrumentation cannot be discussed independently of technology and its applications becomes apparent when one considers the new recombinant DNA technology that is becoming important in clinical laboratory practice. The techniques of electrophoresis and blot transfer discussed earlier are an integral part of recombinant DNA technology. Because DNA probe technology incorporates some of the current laboratory techniques, it deserves a brief discussion.

One must begin with the DNA molecule in which two chains of the DNA helix are held together by complementary base pairing. The base adenine on one chain forms two hydrogen bonds with its complementary base thymine on the opposite strand of the helix. The base guanine on one chain forms three hydrogen bonds with its complementary base cytosine on the opposite chain or strand of the helix. The bases adenine and guanine belong to the class of nitrogen bases called purines. The bases cytosine and thymine belong to the category of nitrogen bases called pyrimidines. DNA, the component of chromosomes in the nucleus, is thus made up of helical chains held together by complementary base pairing. The bases are linked to deoxyribose and phosphate, which form the backbone repeating unit of deoxyribonucleic acid or DNA molecule. The term nucleotide refers to a single base linked with deoxyribose and phosphate groups. In the DNA molecule, however, there are many nucleotides (polynucleotides) with the deoxyribose-phosphate repeating unit or backbone on the outside of the DNA molecule, and the bases on the inside pairing with the complementary base on the opposite strand of the helix.

Gene Probe Cloning

Cleaving DNA

The large double-stranded DNA molecule must be broken down into small molecules if any discrepancy that may account for the mutation and the resulting disease state is to be found in the base sequence of a codon. Alternatively, viral genomes need to be fragmented and understood before probes can be designed to detect them and the presence of specific viral infection. The cleaving of double-stranded fragments four to six bases long has been made possible with the use of restriction endonucleases.[1] Because these enzymes cleave the DNA molecule at specific base sequences, the various small fragments of DNA molecules, and with them, specific nucleotide sequences, can be isolated by the choice of an appropriate restriction endonuclease. These fragments can be separated by electrophoresis on agarose gel. Over 200 restriction endonucleases have been isolated from bacterial sources.

A restriction endonuclease Hind II isolated from the bacterium *Hemophilus influenzae* cleaves DNA molecules at the center of their recognition sites to produce fragments that do not stick together. In contrast, the restriction endonuclease EcoRI isolated from *E. coli* cleaves away from the center of its recognition site leaving sticky ends. The sticky or cohesive ends of a fragment can be joined or annealed by the use of an enzyme called DNA ligase with another complementary sticky end produced by restriction endonuclease cleavage. Thus a recombinant DNA molecule can be produced. Figure 19–1 illustrates the cleavage effected by EcoRI restriction endonuclease to produce sticky-ended fragments. The figure also illustrates how a sticky end of one fragment can anneal with another sticky end, produced by EcoRI cleavage, in the presence of DNA ligase to form recombinant DNA.

A single-stranded DNA molecule can serve as a template through complementary base pairing to another DNA molecule, a process call replication. Alternatively, the single-stranded DNA molecule can be a template to a complementary RNA molecule, a process called transcription. The sugar in the ribonucleic acid (RNA) molecule is ribose, and the pyrimidine base, uracil, forms a complementary base pair with the purine base adenine. In RNA, the complementary base pairing of guanine to cytosine is the same as in the DNA molecule. The genetic code in the DNA molecule is expressed through the RNA molecule, which moves out into the cytoplasm. Thus, in the cytoplasm, messenger RNA (mRNA), as it moves over the surface of the ribosome, forms the template that positions a triplet sequence of nucleotides. These nucleotides code for particular amino acids. This triplet, or codon base, pairs with the complementary base pair (anticodon) on a structurally unique RNA molecule called transfer RNA (tRNA). The specific amino acid expressed through the codon-anticodon pair becomes linked by the aid of a specific enzyme to the end of the tRNA molecule. The amino acids are linked together in peptide linkage, and the polypeptide chain grows as the mRNA molecules move across the ribosomes, with each set of codons succeedingly encoded by the specific anticodons of tRNA. This process of transfer of genetic information in DNA through RNA by which protein is synthesized is called translation. Obviously, if a mutation in the DNA exists, a wrong base is inserted in the codon, a wrong amino acid is incorporated in the protein, and a metabolic disease results. The codon that codes for glutamic acid is GAG (guanine, adenine, guanine). In sickle cell anemia, adenine is replaced by thymine to form the codon GTG, which codes for valine in the sickle cell protein. If portions of sickle cell DNA that contain the GTG codon could be isolated, and if a DNA probe that was complementary to the GTG base sequence existed, the sickle cell DNA could be identified. Herein comes the technology of gene cloning.

Figure 19–1

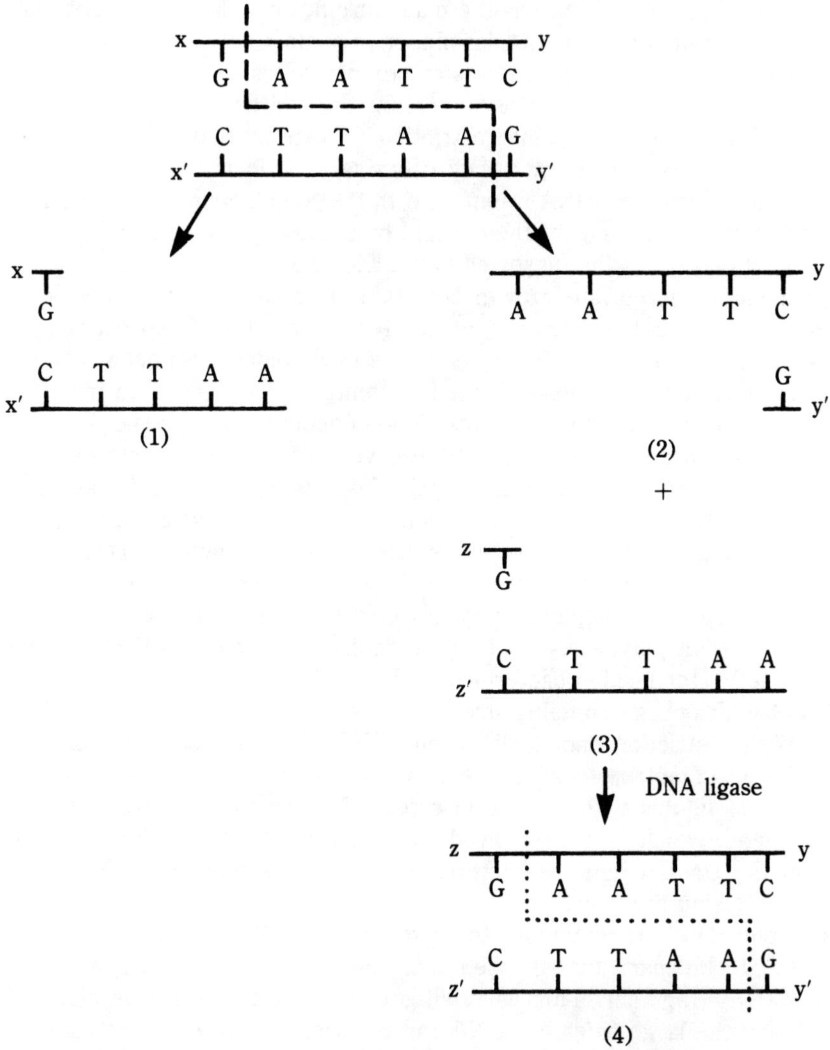

Cleavage of double-stranded DNA by EcoRI restriction endonuclease results in the formation of fragments with sticky ends (1 and 2). Either of these sticky fragments can be annealed in the presence of DNA ligase to a complementary sticky end of another fragment (3) produced by EcoRI cleavage, as is shown with annealed fragment (4).

Cloning DNA

To clone DNA fragments obtained using restriction endonucleases, a vector is needed into which these DNA fragments can be introduced. Bacteria contain, in addition to the main chromosomal material, many small circular extrachromosomal DNA called "plasmids." Whereas the main chromosomal bacterial DNA may have as many as 4 million base pairs, plasmids may contain only several thousand base pairs. Plasmids also carry genes resistant to antibiotics such as kanamycin or tetracyclines. Plasmids can be used as vectors for the desired DNA fragment. Plasmids can be cut open using the same restriction enzyme as that used to prepare the DNA fragment, leaving behind sticky cohesive ends in the region from which a portion was excised. The DNA fragment with the complementary sticky end is then annealed to the sticky ends of the plasmid by use of the enzyme DNA ligase. The recombinant plasmid is then inserted into a bacteria such as *E. coli* and allowed to replicate under appropriate growth conditions in media supplemented with antibiotics. Figure 19–2 illustrates cloning of DNA in plasmid and replication in *E. coli.* In this manner, as many as 200 copies of the recombinant plasmid can be made in replicating bacteria. The cloned DNA fragments can be recovered from the replicated plasmids by using the same restriction endonuclease enzyme that was originally used to excise the plasmid. The recovered cloned DNA fragments can be purified by electrophoresis on agarose gel. Plasmids replicate DNA fragments that are up to 6 kilobases (6,000 base pairs) long. To clone large stretches of a genome, often amounting to 20 kilobases (kb) of DNA or complementary DNA fragments (cDNA, prepared from messenger RNA through the use of enzyme reverse transcriptase to copy the complementary DNA sequence from mRNA and the subsequent use of DNA polymerase to prepare double-stranded cDNA) that may be 45 kb long, the vector used is bacteriophage lambda.

The bacteriophage contains stretches of DNA that are not needed for replication. With restriction endonuclease enzyme, these sites can be excised, and a large DNA fragment up to 20 kb long can be inserted into the space created by excised ends. In this manner, the inserted DNA fragment is packaged into the bacteriophage, which is then allowed to replicate inside *E. coli.* The replicated clones of the DNA within the bacteriophage are subsequently recovered by harvesting and lysing the cells.

To clone cDNA fragments, up to 45 kb long, spaces in the bacteriophage DNA, each 35 to 45 kb apart, the so-called "cos" sites, can be excised to accommodate the long DNA fragment. The "cos" sites of lambda phage can be cloned into a plasmid, and the large fragment DNA can be inserted into the "cos" sites on the plasmid, which is excised with the same restriction endonuclease used to prepare the large DNA fragment. The DNA fragment inserted into the "cos" sites are subsequently cleaved by enzymes that recognize the 35 to 45 kb sequence and are packaged into phage heads. These phage heads then infect *E. coli* for further replication of the inserted DNA fragment. This type of cloning is called "cosmid cloning" because it involves combining "cos" sites of bacteriophage with plasmid.

Preparation or Labelling of DNA Probes

For cloned DNA to be used to identify or probe for complementary base pair sequences in, for example, a viral gene or oncogene, it must first be labelled. The earlier labels utilize the ease with which the phosphate molecules in DNA can be

Figure 19–2

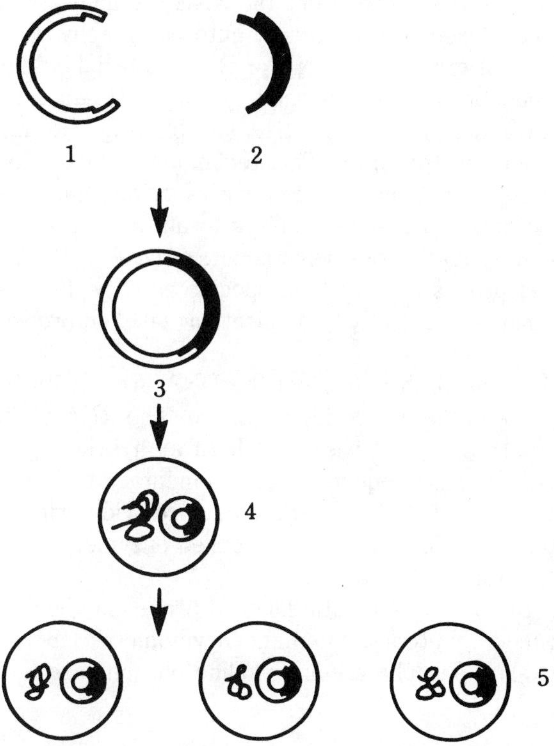

Cloning of DNA in plasmid and replication in bacteria (*E. coli*). 1: Restriction endonuclease–cleaved plasmid with sticky ends. 2: DNA fragment with sticky ends. 3: DNA fragment annealed to plasmid with DNA ligase. 4: Recombinant DNA plasmid inserted into *E. coli.* 5: Replication of plasmid in *E. coli.* Bacterial DNA is indicated by a coil to distinguish it from recombinant plasmid.

radioactively labelled. Thus a DNA probe labelled with ^{32}P can be used in conjunction with the Southern blotting technique discussed in Chapter 17. Briefly, fragments of DNA in the assay sample cleaved by one or several restriction endonuclease enzymes are separated according to their size by agarose gel electrophoresis. The agarose gel is laid over a sheet of nitrocellulose, and buffer is circulated through the gel to allow transfer of separated DNA fragments onto the nitrocellulose sheet to which they become bound. A ^{32}P-labelled probe with a base sequence complementary to the desired DNA fragment to be detected is added to the filter and allowed to hybridize. If the desired DNA base sequence is present in the sample, the probe will bind to it and be visible by autoradiography (indicated by darkened spots on an X-ray film laid over the hybridized probe). Radiolabelled ^{32}p is limited because of its short half-life of 14.3 days, and must be used at a relatively high specific activity to obtain adequate sensitivity. A variation of the Southern blot technique for probing RNA fragments is the northern blot technique, in which the electrophoretically separated fragments are blotted onto diazobenzyloxymethyl paper. If formaldehyde is used to denature the RNA fragment, the fragments can be blotted onto nitrocellulose and detected by autoradiography.

Another method of probing for DNA or RNA material, without prior cleavage with restriction endonuclease and subsequent fragment separation by electrophoresis, relies on extraction of DNA or RNA and blotting onto nitrocellulose filter, or application to slots on the filter. This technique is called dot or slot blotting. The immobilized DNA or RNA is hybridized with specific complementary radioactive probes, and the hybridized probe is visualized by autoradiography. The method can be made sensitive for detection of small amounts of DNA or RNA in sample through use of radiolabelled probes of very high specific activity. Because of the hazards associated with handling radioactivity, nonisotopic labelled probes have been introduced.

One nonisotopic approach is to label the DNA probe with biotin by attaching it to the base thymine in the nucleotide sequence. The RNA probe can be similarly labelled with biotin through the base uracil. After hybridizing the biotin-labelled probe with the sample, base sequences complementary to the labelled probe can be visualized by adding avidin-labelled enzyme, since avidin will bind to biotin. After a washing step to remove unbound avidin-labelled enzyme, substrate is added, and the color of product formed is detected.

Alternatively, the hybridized biotin-labelled probe may be detected by using an antibody to biotin and an appropriate immunoenzymometric label such as peroxidase-antiperoxidase, alkaline phosphatase-antialkaline phosphatase.[2]

Applications

DNA probes are currently applied to detect infectious disease, metabolic diseases, and cancer. DNA probes against viral genomes permit detection of viruses such as herpes simplex types I and II, papilloma virus, adenovirus, Epstein-Barr virus, and hepatitis B virus. Metabolic diseases, such as sickle cell disease, have been detected by using restriction endonucleases that recognize the base sequence of codon in normal DNA (GAG), but not the sickle cell codon resulting from mutation (GTG). The resulting large 1.3 kb fragment from sickle cell gene can thus be distinguished from the smaller 1.1 kb fragment from a normal gene.[3] Alternatively, a probe complementary to the point mutation on the DNA molecule can be used. This approach detects not only sickle cell anemia but also β-thalassemia.[4] In cases where the exact genetic sequence related to a disorder is unknown, a technique called "restriction fragment length polymorphism" (RFLP) can be used. This technique detects variation in the length of fragments obtained by restriction endonuclease treatment. Such RFLP linkage studies may uncover the right probe to use to detect the gene sequence responsible for disease, as is the case with Huntington's disease.[5]

Because several diseases that cause cancer are associated with chromosomal mutations, use of gene probes can aid the understanding of the location of chromosomal aberrations. Because there is an association between viruses and cancer, DNA probes for viruses can be used to better understand this phenomenon. Use of the Epstein-Barr virus DNA probe has revealed an association of this virus and lymphoma.[6] Recombinant DNA methods have shown an association between human T-cell leukemia-lymphoma virus type I (HTLV-I) and a subtype of T-cell leukemia.[6]

Newer amplification techniques permit use of sample without prior extraction of DNA and have simplified prenatal diagnosis of genetic diseases.[7] Amplification of a specific DNA sequence by a factor of 10^6 can now be achieved by the polymerase chain reaction (PCR). In this technique a particular sequence of DNA can be amplified severalfold by repeated cycles of DNA denaturation, annealing to primer with their complementary sequences, and extension of DNA synthesis with DNA polymerase. The amplified viral DNA can be detected by hybridization to a ^{32}p-labelled DNA probe complementary to a portion of the amplified gene segment.[8] Considering that fewer than one peripheral blood lymphocyte expressing viral protein may be found in 10,000 cells of an infected individual, the sensitive PCR technique can provide results within 3 days, whereas viral culture from such a small sample takes from 3 to 4 weeks.

References

1. Watson D, Tooze J, Kurtz DT: Recombinant DNA: A Short Course. Scientific American Books, WH Freeman, New York, 1983.
2. Sauls CD, Caskey CT: Applications of recombinant DNA to pathologic diagnosis. Clin Chem 31:804, 1985.
3. Chang JD, Kan YW: A sensitive new prenatal test for sickle cell anemia. N Eng J Med 307:30, 1982.
4. Orkin SH, Markham AF, Kazazian HH, Jr: Direct detection of the common mediterranean β thalassemia gene with synthetic DNA probes: An alternate new approach for prenatal diagnosis. J Clin Invest 71:755, 1983.
5. Gusella JF, Wexler NS, Conneally PM, et al: A polymorphic DNA marker genetically linked to Huntington's disease. Nature 306:234, 1983.
6. Manzari V, Agliano AM, Gallo RC, Wong-Staal F: A rapid and sensitive assay for proviral sequences of a human retrovirus (HTLV) in leukemia cells. Leuk Res 7:681, 1983.
7. Kogan SC, Doherty M, Gitschier J: An improved method for prenatal diagnosis of genetic diseases by analysis of amplified DNA sequences: Application to Hemophilia A. N Engl J Med 317:985, 1987.
8. Ou C, Kwok S, Mitchell SW, et al: DNA amplification for direct detection of HIV-1 in DNA of peripheral blood mononuclear cells. Science 239:295, 1988.

20
Reflectance Measurements

When a surface reflects light as a mirror does, the reflectance is called specular reflectance;[1] the angle of incidence of light striking the surface is equal to the angle of reflection. Alternatively, if a dull or unpolished surface partially absorbs and scatters in all directions the remainder of the light incident on it, the reflectance is referred to as diffuse reflectance. A matte or perfectly diffusing surface is ideal for reflectance measurements, because it is diffuse reflectance that is measured by reflectance densitometers. A material such as magnesium carbonate or barium sulfate that diffusely reflects all of the light incident on it has a reflectance value approaching 1 (reflectance is the ratio of the amount of diffuse reflectance to the amount of incident radiation intensity). These materials can thus be used as calibration standards. Reflectance measurements at the wavelength of diffuse reflectance are performed relative to diffuse reflectance from a standard. Results are commonly expressed as percent reflectance (%R), which is the ratio of sample reflectance (R_s) to reflectance of reference (R_r) multiplied by percent reflectivity of the reference, which is usually assumed to be 100. Thus,

$$\%R = \frac{R_s}{R_r} \times 100$$

The relationship between reflectance and concentration is nonlinear. Algorithms based on equations devised by Kubelka and Munk in 1931, and Williams and Clapper in 1953, are generally used to linearize the reflectance concentration plot.

Instrumentation

In general, light from a source such as a tungsten, tungsten-halide, quartz-halide, or xenon lamp is collimated and strikes the sample. The angle of the incidence

Figure 20–1

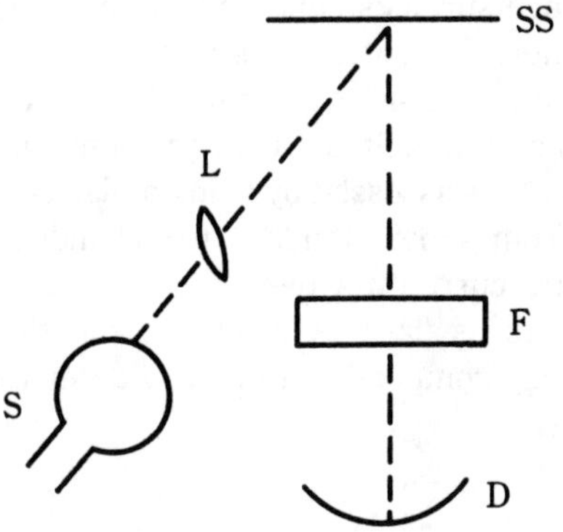

Schematic representation of a reflectance densitometer. S = light source; L = collimating lens, SS = sample surface; F = filter of appropriate wavelength; D = detector.

radiation is presented to the sample and the angle at which diffuse reflectance is measured varies according to the instrument design. Thus light from source as illustrated in Figure 20–1 may strike the sample at an angle of 45°, and diffuse reflectance from the sample isolated by a filter corresponding to the wavelength of reflected light is measured at an angle of 90° to the sample surface. Phototubes or photomultiplier tubes may serve as detectors. Alternatively, the incident light from the light source can be directed at the sample surface at an angle of 90° and diffuse reflectance isolated by a filter of appropriate wavelength measured at an angle of 45°. Some instruments present the sample surface the appropriate wavelength of absorption and reflectance by utilizing a filter in the path of the incident beam.

Instruments using a monochromatic source such as a light-emitting diode can obviously dispense with a dispersing device such as a filter, since monochromatic radiation of the same wavelength as the sample absorption and reflectance wavelength is presented to the sample surface.

An efficient way to measure reflectance is to use an integrating sphere. The integrating sphere is coated internally with either barium sulfate or magnesium oxide or an appropriate efficient diffuse reflector such as a high-efficiency white reflectance paint. In the integrating sphere arrangement, light from a source lamp such as tungsten halide or xenon lamp passes through a dispersing device (filter, monochromator) prior to striking the sample surface. Diffuse reflectance from the surface is captured by the integrating sphere, which is then detected by the photomultiplier tube. Because the integrating sphere itself is a highly efficient diffuse reflector, it serves to collect and modulate the reflectance from the sample surface. The instrument design could be such that the light beam isolated by the dispersing device can be split with a chopper to fall alternately on both the sample and a white reference surface coated with an efficient diffuse reflector such as barium sulfate. In this arrangement the photomultiplier tube can detect diffuse reflectance from both sample and reference surfaces collected by the integrating sphere.

The integrating sphere is central to many of the rapid scanning reflectance densitometers.[2] Reflectance densitometers need to be calibrated with an efficient diffuse reflector such as barium sulfate or magnesium carbonate, or an appropriate white substance. Stray light is assessed by using a black or 0% reflectance standard. A reflectance reading from a 0% standard would indicate the presence of stray light, photomultiplier dark current, or even specular reflectance from the 0% standard, which is referred to as flare. A reading from the 0% standard should be substrated or electronically compensated to allow determination of true reflectance from the sample surface.

Applications

Hand-held reflectance densitometers have made it possible for urine and serum reagent dipstick measurements to be performed in the privacy of one's home. Discrete reflectance analyzers have fulfilled the need for dry chemistry analysis.[3-6] Reflectance measurements on the reagent strip may be made on the same surface as the one to which sample is applied. Alternatively, where the drug chemistry format uses multiple layers of films, the light reflected by the reacted sample is directed onto the reverse side of the spreading layer (the layer that originally received the sample), which is coated with a reflecting material such as titanium dioxide or barium sulfate. The reflectance layer reflects back the diffuse reflectance from the sample onto the detector. In this arrangement, the diffuse reflectance from the sample is read from the side opposite to the side that received the sample.[4]

The development of automated computerized reflectance instruments for urine analysis with the ability to measure reflectance at multiple wavelengths has permitted the measurement of as many as nine parameters on a single reagent strip. Measurements at each of the nine zones may be made by using discrete filters of the appropriate wavelength corresponding to the color of the zone. Alternatively, a separate set of reflectometer and detector combinations may be used for each zone with the aid of light-emitting diode-detector array. In the latter arrangement, two additional light-emitting diode-detectors can correct for reflectance due to urine color and also ensure that the reagent strip is properly aligned for reflectance measurements.

References

1. Kortum G: Reflectance Spectroscopy: Principles, Methods, Applications. New York, Springer-Verlag, 1969.
2. Genshaw MA, Rogers RW: Rapid Scanning Reflectance Spectrophotometer. Anal Chem 53:1949, 1981.
3. Walton B: Dry reagent chemistries in clinical analysis. Anal Chem 55:498A, 1983.
4. Shirey TL: Development of a layered-coating technology for clinical chemistry. Clin Biochem 16:147, 1983.

5. Steinhausen RL, Price CP: Principles and practice of dry chemistry systems. In Price CP, Alberti KGMM (eds): Recent Advances in Clinical Biochemistry, vol 3. Edinburgh, Churchill Livingstone, pp 273–296, 1985.
6. Chu AH, Lopatin W: Application of pattern-recognition techniques in wavelength selection for instrumentally read reagent strips. Clin Chem 32:1666, 1986.

Appendix

Approaches to Laboratory Automation

Currently, clinical laboratories utilize a variety of automated analyzers. Because of the proliferation of such analyzers, singling out some of the analyzers by their trade name might be construed as endorsements. This discussion, therefore, will be limited to approaches to laboratory automated instrumentation.

Steps in analyses that can be automated include:

- Specimen preparation
- Specimen identification
- Accurately aliquoting appropriate specimen volume and reagents
- Exclusion of proteins by dialysis where necessary
- Mixing specimen and reagents
- Incubating for precise time and at proper temperature to effect reaction
- Measurement of reaction
- Reporting of results

Little consideration has been given to automating basic specimen preparation, eg, centrifugation and aliquoting of serum or plasma for analysis. The potential for automation, however, exists with the use of robots.

Specimen identification can be accomplished with some of the current instruments because of their ability to generate coded labels and to decipher the labels with bar code readers.

For many of the automated analyzers in use today, automation begins with the step of aspiration of the prepared specimen and ends in the printout of analytical results. Two main approaches to automated analysis are:

- Analysis in a continuous flowing stream
- Analysis of each specimen discretely

The former is the continuous-flow analysis approach, while the latter is discrete sample analysis.

Continuous - Flow Analysis

The dominance of continuous-flow analysis in many clinical laboratories is the result of the pioneering work of Dr. Leonard Skeggs. The bubble, which is the key to continuous-flow analysis, maintains the integrity of the sample in a flowing stream.[1-5] This is accomplished during the step in which the sample probe rises to aspirate water between each sample. As the sample probe lifts after aspirating the sample, and prior to entering the water receptacle, it introduces an air bubble. More than one air bubble may be introduced if the probe dips more than once into the water receptacle, as is the case with some of the new generation of computerized continuous-flow analyzers.

In addition to the bubbles introduced through the sampling probe, additional bubbles are introduced through a separate tubing that is under an air-release bar, so that each sample-reaction mixture is segmented by a string of bubbles.[2] As each sample-reagent segment is pumped through non-wettable plastic tubes by a peristaltic pump, the air bubbles scrub the interior of the tubing, thus minimizing sample carryover. The segmented air bubbles also ensure that the sample-reagent mixture tumbles through mixing coils, effecting efficient mixing. Carryover between samples can be minimized by including sufficient amount of water wash between samples.

The exact volume of sample and reagent pumped through continuous-flow analyzers is dependent on the internal diameter of the pump tubing. As such, if the internal diameter of the tubing changes with time, the ratio of sample and reagents pumped through also changes and influences the analytical results.

The dialyzers used to remove proteins in continuous-flow analyzers have progressed from early bulky dialyzers. The early dialyzers had a groove length of 88 inches, developed into systems with 6 to 12 inch groove lengths, and are currently miniaturized to as low as 1 inch in computerized continuous-flow analyzers. The newer systems also use a thinner dialysis membrane for effective dialysis. The sample stream dialyzes into a recipient stream (co-current dialysis) at a controlled temperature.

The heating baths used to heat reaction mixtures have also undergone a considerable degree of miniaturization in current computerized systems. The same is true of the colorimeter flow cell. Flow cell volumes have decreased from 3 to 4 mL/minute to 1 mL/minute to as low as 2 μL. Likewise, removal of air bubbles from the flow cell prior to entrance into the colorimeter is no longer necessary with some of the computerized systems that ignore these bubbles.

The direction of radiant energy emerging from the cuvette flow cell to the detector can also be rendered efficient by the use of fiber optics.[3–5] These optical fibers may be made of either glass or plastic. To transmit UV light, fibers made of quartz are used. A material of low refractive index (cladding material) is used to coat the optical fiber, allowing it to transmit light by total internal reflection. Whereas a single optical fiber is limited to transmitting light passing through it, a bundle of optical fibers has the ability to transmit images.

Some computerized continuous-flow systems can perform a simultaneous analysis of as many as 20 serum constituents at a rate of 150 samples per hour, using only 600 μL of serum.[4,5] The computerized systems have the versatility ranging from correcting for sample carryover, to detecting problems and alerting the operator to problems and possible remedial steps, to applying quality control rules to acceptance or rejection of data.[3–5]

Discrete Sample Analysis

In discrete sample analysis, each sample is handled separately, in contrast to continuous-flow analysis, in which samples are analyzed in a flowing stream that is segmented by air bubbles. In effect, in discrete sample analysis, each sample has its own reaction cuvette.

There are various approaches to discrete sample analysis. The analysis can be performed in a "batch" mode, ie, a single test can be performed on a series of specimens in an analytical run. The centrifugal analyzer is an example of a single test batch analyzer. In a typical centrifugal analyzer, transfer discs containing sample and reagent are loaded on the rotor by an automated pipetting station. Cuvette rotors with 15 to 42 cuvette wells are available. Initially, sample and reagent are kept separate in their respective wells. As the rotor is accelerated by application of centrifugal force, mixing of the sample and the reagent occurs, and the absorbance of the mixture in the respective cuvette wells is monitored as each well passes across a stationary optical system at speeds of up to 5,000 RPM. It is possible to take measurements within 2 to 5 seconds of accelerating the rotor and effecting mixing, and to make subsequent measurements during each revolution of the rotor. Assuming 100 milliseconds are required for one complete revolution of the rotor, ten absorbance measurements will be taken from each cuvette well or sample every second. For a 15-cuvette rotor containing 14 samples and a reference sample, this allows a total of 150 absorbance readings each second. For this reason, centrifugal analyzers are interfaced with minicomputers to handle the enormous amount of data.[6,7]

Because multiple measurements are taken on all samples in the analytical run almost simultaneously during each ultrarapid revolution of the rotor, analysis performed by centrifugal analyzers is also referred to as multiple parallel analysis. Centrifugal analyzers are ideally suited for rate measurements of enzyme activity or other kinetic measurements of analyte reactions. Some centrifugal analyzers also have the capability to make fluorescence and nephelometric measurements.[7]

Noncentrifugal batch analyzers are also available. Some of these analyzers make measurements at two different wavelengths, the analytical wavelength and an interferent wavelength. In this bichromatic approach, any absorbance at the interferent wavelength is automatically subtracted from the absorbance at the analytical wavelength.[8,9]

Discrete sample analyzers may also use a sequential discretionary approach. In such analyzers, specimens are presented sequentially for analysis with information regarding the test required on the specimen entered on the keyboard. Analysis is performed in the order that the specimens are presented to the instrument. Such sequential discretionary analysis can be performed in a distinct reaction cup. Reagent is first dispensed to cover the cell, and the excess reagent is siphoned off, a procedure which allows the same amount of reagent to be dispensed each time. The sample is then added, mixed, and measurements are made. Measurements of either absorbance, potential (ion-selective electrode measurements) or current (amperometric or polarographic measurements) can be made with such analyzers. After the measurement on a sample is complete, the reagent mixture is drained out, and the reaction cup is washed with fresh reagent to eliminate carryover for the next sample. The pipette tip is also washed between samples.[10-12]

Another approach to discrete sample analysis is random access.[13,14] In the random access analyzer, the tests required on a particular specimen are completed before moving onto the next specimen. A stat specimen may also be introduced to supersede other specimens. The microprocessor determines the sample volume, the appropriate reagent, and its volume and the wavelength or other analytical parameter required for measurements. Currently, random access analyzers are in favor.

Many approaches are possible in random access analysis. An analyzer may use a plastic transparent package that contains tableted reagents. The reagents are crushed and mixed with the serum and diluent inserted previously into the pack. The package subsequently moves to the photometer area where the bag is crushed to form a plastic cuvette through which radiant energy from a lamp passes. The appropriate filter wheel is chosen to make absorbance measurements.[15] Alternatively, random access analyzers may use reagents that are compartmentalized in layers of film on a slide.[16] In such multifilm layered analysis, sample that is pipetted onto the film diffuses into reagent layers and triggers the reaction. Reflectance measurements are made, although ion selective electrode measurements are also possible for slides designed to measure electrolytes. The compartmentation of reagents in separate film layers allows the use of incompatible reagents, since these reagents will be admitted only when the sample is added. It is possible to perform rate measurements by the use of a second reflectometer that measures the rate of change in reflectance at intervals of a few seconds.

Both continuous-flow and discrete sample analyzers have increased sample throughput, ie, the maximum number of samples or tests that can be analyzed in an hour. They have also made possible the rapid measurement of enzyme activity by allowing rate measurements of initial velocity. For samples containing normal enzyme activity, the absorbance change will be very low. In contrast, samples containing high enzyme activity can be measured rapidly before substrate depletion occurs. In fact, the optimum range of absorbance measurements is dependent on the photomultiplier noise and the efficiency of the analog to digital (A/D) converter, which translates the photomultiplier signal to a digital readout. Efficiency of the latter is defined by the extent of resolution of a full scale signal. For example, a 12-bit (2^{12} = 4,096 counts) A/D converter used with currently available centrifugal analyzers can resolve a 0–2 absorbance signal up to 0.0005 absorbance/count (Note: 2/4,096 = 0.0005). Thus, sensitivity to levels of 10^{-4} absorbance units over a very short interval of 10 seconds can be achieved in some of the discrete sample analyzers utilized for rate assays.[17]

Computerized continuous-flow analyzers compare the reaction slopes of flow streams corresponding to the three time intervals at which they reach the flow cells. The enzyme activity value is reported only if there is a constancy of slope between the first and the second flow streams, and the first and the third flow streams.[17]

Many of the discrete sample analyzers monitor reaction slopes at either preset time intervals or during a short time interval, such as a fixed 17.1-second interval.[17]

In addition to monitoring the rate (da/dt) or the first derivative for a preset time, such as 17 seconds, the second derivative (d^2a/dt^2) is also monitored. If the rate is constant, the second derivative by definition is zero. Thus, the recorder pen monitoring the second derivative will stay on the baseline, thus assuring that no substrate depletion has occurred and that the enzyme activity reported is the true value.[17]

Current and Future Perspectives

The performance characteristics of selected instruments currently in use are discussed in a variety of sources,[18-26] including books dedicated to describing and cataloging automated instruments.[27,28] Some factors contributing to the efficiency of automated analysis need to be highlighted.

Sample and Reagent Delivery

The precision of results obtained with automated analysis rests on the ability to pipet microliter amounts of sample and reagent. Microprocessor controlled, step motor–driven pumps that electronically regulate sampling of exact volumes of sample and reagent with great precision are available.[27] Varying the number of voltage pulses supplied to the step motor allows the volume pipetted to be changed rapidly, permitting within the same run the analysis of analytes requiring different sample-to-reagent ratios. Sample pickup and delivery into the reaction cell in discrete analyzers is generally accomplished with step motor–driven syringes that operate by either the air displacement or positive displacement principle. In the air displacement approach, the sample tubing connected to the syringe pump draws the sample from the sample receptacle through the suctional force of air in the syringe as the plunger is pulled under step motor control. It is later deposited into the reaction cell with a flush of fluid that serves to wash the contents of the sample probe into the reaction cell. In this arrangement, an air bubble can also be introduced prior to sampling to prevent the sample from coming into direct contact with the fluid in the sample tubing. Furthermore, the introduction of an additional bubble after some of the sample has been withdrawn allows the first air bubble sample slug to scrub and clean the sample tubing wall, which is then discarded, while the second air bubble-sampling slug is deposited into the reaction cell for analysis. The sample probe is subsequently cleaned either by dipping it into a deionized water receptacle or by spraying water on the probes, which effectively cleanses both the inside and outside of the sample probe, and thus minimizes carryover into the next sample.

Other novel approaches have been attempted to insure precise metering of fluids. The use of precision tooled plastic blocks with channels to effect mixing of sample and reagent provides an elegant tubing-free solid state fluid handling system requiring minimum preventive maintenance.

Minimizing Carryover

To insure accuracy in automated analysis it is important to minimize carryover from the previous sample. Washing by dipping the sample probe into a washing receptacle or spraying the sample probe with water is an approach already described under positive displacement sample pick-up and transfer. Coating of the sample probe with a nonwettable material such as Teflon can minimize carryover. Limiting the depth of insertion of the probe into the aspirating fluid receptacle reduces carryover from outside the probe.

Limiting the probe's depth of insertion can be accomplished with liquid-level monitoring systems. Liquid level can be sensed in a variety of ways. Resistance detectors measure the difference between low conductivity while the probe is in air, and high conductivity just as the probe hits the liquid level.[18] Optical sensors measure the difference in transmission of visible light or infra red light between air and liquid, which in effect can adjust the sample tube height to the minimal level required to aspirate the sample.[8] Alternatively, wave detectors sense liquid level by introducing a wave and following the resulting oscillations. Monitoring change in capacitance is another approach for detecting liquid level. The probes themselves can be equipped with light-sensitive diodes to insure that required aliquots have been aspirated. Finally, liquid level sensing systems can also alert the operator if the reagent level is nearing depletion.

Carryover is considered to be a greater problem with continuous-flow than with discrete sample analyzers. However, currently used random access continuous flow based analyzers have been adapted to greatly minimize carryover. One approach is to use an inert liquid fluorocarbon, which is nonwettable by nature, to coat both the inside and outside of the probe to prevent carryover.[13,14] An approach to virtually eliminating carryover in slide-dry chemistry analyzers is by using a disposable plastic tip for each sample that is aspirated.

Another approach is to carry out the reaction in a Teflon tubing. Here the sample and a reagent separated by small air bubbles and another reagent are encapsulated in the liquid fluorocarbon to minimize carryover. Each encapsulated sample is separated by the previous or succeeding sample by a slug of buffer and large air bubbles. As each encapsulated sample reaches the position of the Teflon tubing where the tubing diameter broadens, the small air bubbles in the sample capsule separating sample, and reagents from the other reagent collapse. They merge with the larger air bubbles and effect mixing of sample and reagents to initiate reaction. The Teflon tubing containing the reaction mixture also serves as a cuvette through which photometric readings are taken. In this continuous flow random access system, the use of a nonwettable Teflon tubing and the non-wettable fluorocarbon to encapsulate sample and reagents effectively minimizes carryover.[19]

Sample Identification

With the hierarchy of automated analyzers currently available (ranging from analyzers that can perform greater than 2000 tests per hour, to analyzers that perform between 600 to 1200 tests per hour, to analyzers that can perform as few as 300 tests per hour or less), maintaining sample identification and keeping sample manipulations to a minimum become challenging tasks, especially with high-throughput, highly mechanized analyzers.

Ideally, sampling directly from the primary blood collection tube to which a bar code has been affixed should provide the best form of positive sample identification. Sampling directly from the primary collection tube with the sample probe piercing the tube stopper is featured in some of the latest hematology and chemistry analyzers, a trend that will accelerate because of the currently heightened awareness of reducing infectious hazards in the clinical laboratory. Ideally, the bar code should be the same as the patient identification code and should be affixed to the primary specimen container at the time of specimen collection, but this rarely occurs. Laboratories using the bar code usually affix the bar code to the specimen in the laboratory. Furthermore, laboratories that pour off serum or plasma to sample cups can maintain positive identification only by the position of the sample cups in the sample tray. This can be accomplished in some instruments by the use of binary-coded sample carousels or by the sequence of placement of samples as determined by the instrument.

The use of bar codes, attractive as it may seem, is not without problems. There is a proliferation of bar codes due to the absence of a standard bar code, though progress is being made in this area, and a consensus toward a standard bar code with 43 data characters (Code 39) may be on the horizon.[27]

Bar codes should be printed on appropriate reflective index paper to permit them to be scanned by the instrument by means of a laser scanner. Because a bar code is a combination of bars, spaces, and characters, any unwanted marks or smudges caused by dirt or blood spots in the clear areas of the bar code could lead to misreading by the scanner. Sometimes improper positioning of the bar code on the tube can impede proper identification. Fortunately, there are built-in safeguards to minimize misreading by the scanner, such as the use of check digit characters in the bar code to match and verify the identification number on the bar code.

In addition to providing positive sample identification of the primary sampling tube, the bar code serves other useful functions that collectively increase the efficiency of automated analysis. Operator entry of tests required on the specimen could be dispensed with if the instrument, after reading the bar code, has the ability to interact with the host computer to decipher and perform the tests requested on the specimen. Bar codes are also helpful in identifying reagents and alerting the operator if they have not been placed in their usual position, or if a reagent has reached its expiration date.

Efficiency of High-Speed Automated Analysis

The efficiency of high-speed automated analysis ultimately depends on its ability to orchestrate the various microprocessor controlled functions of data collection, data storage, rapid access, the ability to monitor the functioning of the various modules by detecting calibration drifts, and the ability to automatically recalibrate the instrument when calibration limits are exceeded. The throughput of an analyzer depends on analytical and sampling cycle time. The analytical cycle is governed by the rate at which a test can be initiated. An instrument with a 5-second analytic cycle can perform 720 tests per hour. The sample cycle is the rate at which a new specimen can be accessed by the analyzer.

Role of Robotics

Analyzing the greatest number of samples in the shortest possible time, otherwise known as tandem-access scheduling, is a challenge that can be accomplished only by great speed of delivery and manipulation of samples fed to the instrument. In this context, robotics function and represent the future of automation. One laboratory that has efficiently used robotics is the laboratory of Professor Masahide Sasaki at the Kochi Medical School in Japan. His laboratory uses a conveyor belt system to transport the samples to the automated analyzers which are in turn connected to the conveyor belt. Robotic analyzers are in use in Dr. Sasaki's laboratory to perform all the steps associated with serological aggregation tests, including AIDS tests. He has also used robotic analyzers to automate all the functions related to blood transfusion tests, such as ABO blood typing, crossmatching, and Rh factor testing. Robots also use designated tracks in the Kochi laboratory to carry out simple tasks of transportation of materials from one location to the other.

Robots of variable flexibility, ranging from three degrees of freedom (cartesian), to four degrees of freedom (cylindrical), to the highly flexible five degrees of freedom (jointed), have been used to perform a variety of tasks, such as placing pipette tips on pipetting probe, gripping tubes, mixing liquids, and conducting extraction and sample preparation steps.[28] Because the movements of the robotic arms are under the control of complex electronic circuits, any fluctuation in line voltage can disrupt the operation of the robot. As such, an uninterrupted filtered power supply with back-up batteries is an important requisite for laboratories using robots.

With robotics, many of the repetitive functions currently performed manually, such as sample preparation, including centrifugation and sample aliquoting steps, and monitoring calibration and performing preventive maintenance on instruments, will be replaced.

Image Analysis

Current developments in urine analysis automation and image analysis of urinary sediment are encouraging. Specific gravity of urine can be determined by detecting change in frequency of sound wave introduced into the urine. Urinary sediment can be examined on a moving specimen instead of on a traditional slide by freezing the flow of urine with bursts of light and taking several video pictures of the specimen under a microscope.[28]

Image analysis has been applied to scanning of two-dimensional electrophoresis gels.

Emerging Technology

Technology such as time-resolved fluorescence[29] adapted to immunoassays have been incorporated in dedicated time-resolved fluorometers and are attractive candidates for large scale automation. The availability of lanthanide chelates (Europium chelates) whose excitation and fluorescence emission spectra are as much as 290 nm apart, as opposed to the 28 nm separation for fluorescein, makes the Europium chelates attractive probes. Their excitation and fluorescence emission wavelengths can be resolved easily without overlap. The most attractive feature of these Europium chelates is their long-lived fluorescence lifetime, which is of the order of 10 to 1000 microseconds, as opposed to nanoseconds for conventional fluorescent probes. This property is amenable to automation under microprocessor control. The sample can be excited by a short flash of light from a source such as a xenon lamp. After a time delay, during which short-lived background fluorescence fades, a microprocessor can activate fluorescence wavelength selector to take measurements.

Many new technologies now on the horizon such as Biosensors[30] will conceivably be utilized in automated analyses. It is not farfetched to expect plant materials as biosensors. The outer layer of flower petals, although hydrophobic, permit gas exchange. For this reason, flower petals such as those of magnolias or carnations can be incorporated in the tip of an electrochemical sensing element to provide a biosensor. Plant tissue, such as banana, which is sensitive to dopamine, can be used as a dopamine biosensor. Likewise, cucumber can be used as a vitamin-C sensor, onion or garlic as sulfoxide sensors, and the mesocarp portion of yellow squash as a glutamic acid sensor.

References

1. Coakley WA: Handbook of Automated Analysis: Continuous Flow Technique. New York, Marcel Dekker, 1981.
2. Winter SD, Kerr C, McAvoy M, et al: "Carryback": Effect of viscous liquid controls on the preceding sample analyzed with the SMA II continuous-flow analyzer. Clin Chem 31:1896, 1985.
3. Walmsley TA, Fowler RT, Abernethy MH: Microsampling on the Technicon SMAC system. Clin Chem 29:1531, 1983.
4. Gomez P, Coca C, Vargas C, Acebillo J, Martinez A: Normal reference intervals for 20 biochemical variables in healthy infants, children and adolescents. Clin Chem 30:407, 1984.
5. Robertson EA, Lsako G, Faust A, Elin RJ: Effect of calibration material on lot changes on results for the "SMAC". Clin Chem 30:1402, 1984.

6. Tiffany TO: Centrifugal fast analyzers in clinical laboratory analysis. Crit Rev Clin Lab Sci 5:129, 1974.
7. Hills LP, Tiffany TD: Comparison of turbidimetric and light scattering measurements of immunoglobulins by use of a centrifugal analyzer with absorbance and fluorescence/light-scattering optics. Clin Chem 26:1459, 1980.
8. Driscoll RC, Edwards RB, Liston MD, et al: Discrete automated chemistry system with tableted reagents. Clin Chem 29:1609, 1983.
9. Schultz SG, Holen JT, Donohue JP, Francoeur TA: Two-dimensional centrifugation for desk-top clinical chemistry. Clin Chem 31:1457, 1985.
10. Blumenfeld TA, Griffith B: Measurement of urinary constituents with the ASTRA-8 multichannel analyzer. Clin Chem 26:1883, 1980.
11. Kessner A, Burnett RW, Bowers Jr GN: Response monitoring and action limits: use of ADC numbers in understanding the operational characteristics of the Beckman ASTRA chemistry system. Clin Chem 27:1227, 1981.
12. Janssen JW, Verdegaal MCJM, Visser RWJ: Performance of the ASTRA enzyme systems evaluated. Clin Chem 30:1830, 1984.
13. Schwartz MK, Statland BE, Coughlin J, et al: Chemical and clinical evaluation of the Random Access Analyzer "RA-1000." Clin Chem 30:364, 1984.
14. Ng RH, Altaffer M, Ito R, Statland BE: The Technicon RA-1000 evaluated for measuring sodium, potassium, chloride, and carbon dioxide. Clin Chem 31:435, 1985.
15. Schifreen RS, Cembrowski GS, Campbell DC, et al: A quantitative automated immunoassay for fibrinogen/fibrin degradation products. Clin Chem 31:1468, 1985.
16. Von Schenck H, Treichl L, Tilling B, Olsson AG: Laboratory and field evaluation of three desktop instruments for assay of cholesterol and triglyceride. Clin Chem 33:1230, 1987.
17. Narayanan S: An evaluation of methods for monitoring enzyme activity. ACSP Check Sample CC-104, 1977.
18. Kineiko RW, Floering DA, Morrissey M: Laboratory evaluation of the Boehringer Mannheim "Hitachi 705" Automatic Analyzer. Clin Chem 29:688, 1983.
19. Cassaday M, Diebler H, Herron R, et al: Capsule chemistry technology for high-speed clinical chemistry analyses. Clin Chem 31:1453, 1985.
20. Popelka SR, Miller DM, Holen JT, Kelso DM: Fluorescence polarization immunoassay. II: Analyzer for rapid, precise measurement of fluorescence polarization with use of disposable cuvettes. Clin Chem 27:1198, 1981.
21. Jolly ME, Stroupe SD, Schwenzer KS, et al: Fluorescence polarization immunoassay. III: An automated system for therapeutic drug determination. Clin Chem 27:1575, 1981.
22. Li TM, Robertson SP, Crouch TH, et al: Automated fluorometer/photometer system for homogeneous immunoassays. Clin Chem 29:1628, 1983.
23. Degiampietro P, Peheim E, Colombo JP: Laboratory evaluation of the Greiner G-400 discrete selective multichannel analyzer. Clin Chem 31:302, 1985.
24. Moses GC, Lightle GO, Tuckerman JF, Henderson AR: The EPOS automated selective chemistry analyzer evaluated. Clin Chem 32:165, 1986.
25. Astill ME, Johnson LR, Thorne GH: Dual fluorometric/colorimetric detection system for an automated random-access instrument utilizing standard polystyrene test tubes as precision cuvettes. Clin Chem 33:1554, 1987.
26. Cynober L, Morgant G, Morelet L, Giboudeau J: Laboratory evaluation of the kone "progress" discrete analyzer. Clin Chem 33:123, 1987.

27. Lifshitz MS, De Cresce RP: Understanding, selecting, and acquiring clinical laboratory analyzers. New York, Alan R. Liss, 1986.

28. Lifshitz MS, De Cresce RP (eds): Perspectives on clinical laboratory automation, in clinics in *Laboratory Medicine* vol 8, #4, Philadelphia, W. B. Saunders, 1988.

29. Diamandis EP: Immunoassays with time-resolved fluorescence spectroscopy: Principles and applications. Clin Biochem 21:139, 1988.

30. Rechnitz GA: Biosensors: An overview. J Clin Lab Analysis 1:308, 1987.

Index

Numbers in **boldface** refer to pages on which illustrations or tables appear.

T

U

V

W

X

Z